P9-CPX-121

Ways to be a Better FRESHWATER FISHKEEPER

Mary Bailey, Sean Evans, Nick Fletcher, Andy Green,
Peter Hiscock, Pat Lambert and Anna Robinson

FIREFLY BOOKS

Published by Firefly Books Ltd. 2005

First printing

Publisher Cataloging-in-Publication Data (U.S.)
Bailey, Mary.
500 ways to be a better freshwater fishkeeper/Mary Bailey ... [et al.]
[128] p. : col. photos. ; cm.
Summary: Tips for tropical freshwater fishkeepers, including information on aquariums, heating and lighting, water management, aquarium plants, aquascaping, feeding, compatibility, maintenance and healthcare.
ISBN 1-55407-048-1
1. Aquariums. II. Title.
639.34 dc22 SF457.B355 2005

Library and Archives Canada Cataloguing in Publication
500 ways to be a better freshwater fishkeeper / Mary Bailey ... [et al.]
ISBN 1-55407-048-1
1. Cichlids. 2. Aquarium fishes. I. Bailey, Mary II. Title: Five hundred ways to be a better freshwater fishkeeper.
SF457.F58 2005 639.3'774 C2005-902197-7

Published in the United States by
Firefly Books (U.S.) Inc.
P.O. Box 1338, Ellicott Station
Buffalo, New York 14205

Published in Canada by
Firefly Books Ltd.
66 Leek Crescent
Richmond Hill, Ontario L4B 1H1

Created and compiled: Ideas into Print, Claydon, Suffolk IP6 0AB, England
Design and prepress: Stuart Watkinson, Ayelands, New Ash Green, Kent DA3 8JW, England.
Practical photography: Geoff Rogers
Production management: Consortium, Poslingford, Suffolk CO10 8RA, England.
Print production: SNP Leefung

Printed in China

Contents

Better Aquariums

1 Calculating the volume of water your tank will hold

To calculate a rough estimate of the volume of a rectangular or square tank in gallons, multiply the length by the width by the depth in inches and divide by 230. If working in metric, multiply the dimensions in centimeters and divide by 1,000 to estimate the volume in liters. But remember that most tanks will not be filled to the brim and that decor and substrate can reduce effective water volume by up to 30 percent. Many manufacturers clearly indicate the volume of their tanks, and some include it in the product's name. With others, a number in the name has no correlation with the volume, so always check.

2 Be sure to select glass of the correct thickness

Tanks up to 48 x 18 x 12 inches (120 x 45 x 30 cm) can safely be constructed with new $^{1}/_{4}$ inch (6 mm) glass (not salvage or offcuts), providing they include bracing bars. Depth, rather than length or width, determines the outward pressure exerted on the panels. Any tank deeper than 18 inches (45 cm) should be made with $^{1}/_{3}$ inch (10 mm) glass, while the base should be of $^{1}/_{2}$ inch (12 mm) glass in tanks 60 inches (150 cm) or longer, to resist flexing.

3 Standard-sized tanks offer the best value

Thanks to the economies of large-scale production, it is invariably cheaper to purchase a standard-sized tank from a reputable manufacturer than try to build your own. You will then have a product that is durable and good looking; a piece of furniture, in fact, to fit into your home decor. Homemade tanks only really have a place in fish rooms or basements, where the decorative element is not so important.

4 Estimate stocking levels before buying your fish

Tank volume determines how many fish you can keep. An initial stocking guide is 1 inch of fish length per gallon (2.5 cm per 4.5 L), excluding tails. But this applies only to "average-sized" fish 1$^{1}/_{2}$ inches (up to 4 cm) in normally proportioned tanks. Weight-to-length ratio rises significantly after that, and larger fish produce more waste—so a 12-inch (30 cm) oscar *(Astronotus ocellatus)* needs more room than 12 neon tetras *(Paracheirodon innesi)*, measuring 1 inch (2.5 cm) each.

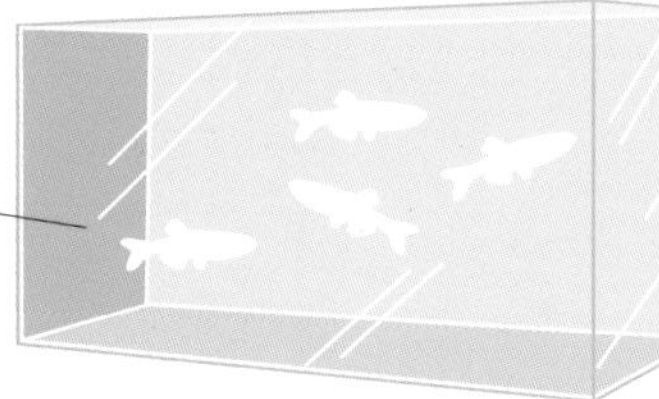

Tip 1 *A 48 x 12 x 12 in. (120 x 30 x 30 cm) tank with four 6-in. (15 cm) fish.*

A 48 x 12 x 12 in. (120 x 30 x 30 cm) tank with 12 2-in. (5 cm) fish.

Sizes and capacities of standard tanks

This table provides a guide to the capacity and weight of water for a range of standard size, all-glass aquariums.

Tank size, L x W x D		Volume of water		Weight of water	
20 x 10 x 12 in.	(50 x 27 x 31 cm)	10 gal.	(38 L)	85 lb.	(38 kg)
24 x 12 x 12 in.	(61 x 31 x 31 cm)	15 gal.	(57 L)	125 lb.	(57 kg)
24 x 12 x 16 in.	(61 x 32 x 41 cm)	20 gal.	(76 L)	165 lb.	(75 kg)
30 x 12 x 16 in.	(76 x 32 x 41 cm)	25 gal.	(95 L)	210 lb.	(95 kg)
36 x 12 x 18 in.	(92 x 32 x 46 cm)	33 gal.	(125 L)	275 lb.	(125 kg)
48 x 16 x 20 in.	(122 x 41 x 51 cm)	66 gal.	(250 L)	550 lb.	(250 kg)

5 Deep fish need deep tanks and tall plants for cover

Because of their body depth, some fish species require deeper tanks than normal. For example, angelfish *(Pterophyllum scalare)* and discus *(Symphysodon aequifasciatus)* are happiest in tanks at least 18 inches (45 cm) wide and preferably 24 inches (60 cm) deep. If you are growing real plants in deep tanks, choose tall species with a low light requirement, such as vals (*Vallisneria* spp.) or anubias, because the fish mentioned do not like their home too brightly lit.

Tip 5
Discus (Symphysodon aequifasciatus) need a deep tank.

6 Choose an acrylic tank for strength

Modern acrylic tanks do not scratch easily and are structurally sounder than tanks made from sheets of glass bonded with silicone sealer. In fact, they are 17 times stronger, a 20 percent better insulator and a fifth of the weight compared to glass. They are a good choice if you are keeping large, boisterous fish such as large cichlids or catfish, which are quite capable of throwing heavy items of decor around.

7 Very tall tanks are not suitable for fishkeeping

Very tall columnar or tubular tanks are fine as futuristic display pieces but are not suitable for keeping fish. Their air–water interface (water surface) is tiny in relation to their capacity, they are difficult to keep clean, and fish cannot behave naturally when their lateral swimming space is so restricted.

8 Is there any advantage to having a bow-fronted tank?

Bow-fronted tanks are visually very attractive, and if your tank has to be sited directly opposite a window the curved glass cuts down annoying reflections, giving you a better view of the fish. Bear in mind, though, that replacing the front panel in the event of breakage will be expensive, and that you cannot clean curved glass with a flat algae magnet!

9 Buy the largest tank possible for a stable setup

Buying the largest aquarium you can afford is a good investment. Besides enabling you to keep more fish, the more water a tank holds, the more stable its environment will be. Any water quality problems will manifest themselves more slowly, and in the event of a power failure the tank will take far longer to cool to the point where the fish are put at risk.

10 Will a tank with a mirrored back stress my fish?

Some small aquariums have a mirrored back glass to give the illusion of more space than they actually provide. These should be avoided because your fish will feel insecure in such surroundings. In any case, algal growth will soon reduce the mirroring effect, and cleaning the back glass is always disruptive for the occupants, since much of the decor will be situated toward the rear of the aquarium.

Tip 12 *Small, systemized tanks are ideal for beginners.*

11 Tanks seem narrower when they are full of water

The refractive properties of water mean that your view from the front to the back of the tank will be shortened by approximately one third. If you enjoy aquascaping, bear this in mind. Another limitation of narrow aquariums is that they have a small base area in relation to volume, not ideal if you keep fish that stake out a breeding territory such as cichlids.

Tip 9 *Tanks of all shapes and sizes are available.*

12 Ideal tanks for budding fishkeepers

Small systemized aquariums of around 5 gallons (20 L) are an ideal and inexpensive way of introducing children to freshwater fishkeeping, and they are light enough when filled to rest safely on a sturdy bedside table. But their limited stocking capabilities are at odds with youthful enthusiasm—5 inches (13 cm) total fish length maximum. This makes them suitable only for small fish such as neon tetras *(Paracheirodon innesi)* and zebra danios *(Brachydanio rerio)*.

13 Is it worth buying a secondhand tank?

Secondhand tanks are an apparently inexpensive route to start fishkeeping, but they carry many risks. Very often they come with equipment and fish you do not want, with all the attendant problems of re-homing, and there is no guarantee to safeguard you if they leak or burst, or if the equipment fails.

14 I would like a fish tank but have very little space

Where space is severely limited, a modern globe tank with built-in filtration and aeration and a capacity of 8 gallons (30 L) is an attractive and viable choice to house a handful of small community fish. However, a basic 24 x 12 x 12 inch (60 x 30 x 30 cm) tank holds 14 gallons (55 L) and is cheaper to buy.

Tip 15 *Hexagonal tanks offer limited lighting options.*

Tip 14 *Globe tanks take up little space.*

15 Hexagonal tanks make an attractive change

A 24-inch (60 cm) tall hexagonal aquarium is a viable alternative to the traditional rectangular tank. It occupies very little floor space and holds approximately 21 gallons (82 L) of water. However, hoods on hexagons have little room for light bulbs so opt instead for a 125-watt mercury vapor lamp, suspended 12 inches (30 cm) above the water surface, particularly if you intend to use natural plants.

16 Systemized tank or package deal?

There is a great difference between "systemized" tanks, with built-in filtration, heating and lighting, and tanks sold as package deals with ancillary equipment merely stacked inside. If the extras are what you would choose to buy anyway, these collections represent good value—if not, it is best to buy a plain tank and add the accessories separately.

Better Aquariums

17 Shop around for the best price

It is well worth comparing prices before you buy an aquarium. There is usually a significant difference between the manufacturer's suggested retail price and what is actually being asked. Discontinued lines can also be a bargain—tanks take up a lot of floor space, and aquatic stores prefer a rapid turnover of stock.

18 Check the access before you buy a large aquarium

Before you order a large aquarium, check that your doors are wide enough to allow it access and that there is sufficient room to maneuver it into its final position. A tape measure will give you some clues, but a dummy run with a cardboard box of the same dimensions will be more reassuring.

19 Play safe and get your aquarium delivered

For all but the smallest aquariums it is worth paying the store a small delivery fee. Getting all-glass tanks home in your car is a perilous exercise unless they will fit on the floor or sideways across the rear passenger seats. And anything longer than a 36-inch (90 cm) tank will be too heavy and cumbersome for one person to maneuver in and out of the vehicle.

20 Inspect new tanks and clean them before use

Inspect your tank carefully in the store, and reject it if it shows scratches or other transit damage. On getting it home, clean the glass inside and out, using only a lint-free cloth and warm water, to rid it of dust. Once it is in position, the first thing to do is fill it and check for leaks—these are rare, but if anything is wrong your dealer should replace the tank.

21 Cushion all-glass tanks to prevent them from cracking

If you have chosen an unframed, all-glass aquarium, you are strongly advised to cushion the tank on its supporting cabinet or stand with a 1-inch (2.5 cm) thick sheet of Styrofoam to iron out any irregularities. The exposed edges of Styrofoam can be easily disguised with black plastic insulating or duct tape. Ceiling tiles are too thin to provide an efficient cushion.

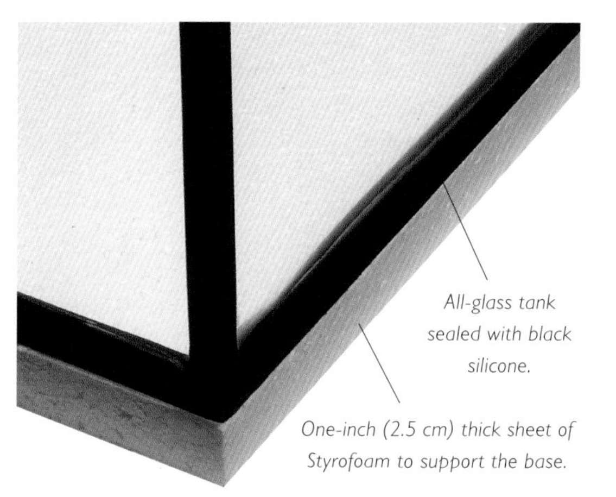

Tip 21 *Provide a generous cushion for all-glass tanks.*

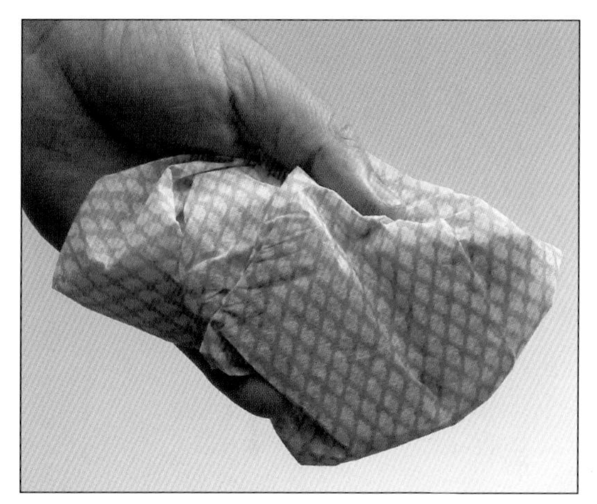

Tip 20 *Clean new tanks with water and a soft cloth.*

22 Tanks with a bottom frame do not need a mat

Some aquariums feature a bottom frame that holds the glass of the base clear of the supporting stand. There is no need here for a cushioning Styrofoam sheet. A further benefit, although not very common, is if the frame is fitted with locating pins that marry up with holes in the stand, making for a very secure, tip-free system.

23 Support heavy tanks on the floor joists

Water is heavy—a 36 x 12 x 12 inch (90 x 30 x 30 cm) tank holds 21 gallons (82 L) weighing 180 pounds (82 kg). Concrete ground floors should support even the largest aquariums with ease, but if the floor is wooden, try to position the stand or cabinet so that the weight is taken on the joists (identifiable by the securing nails) rather than on the floorboards alone.

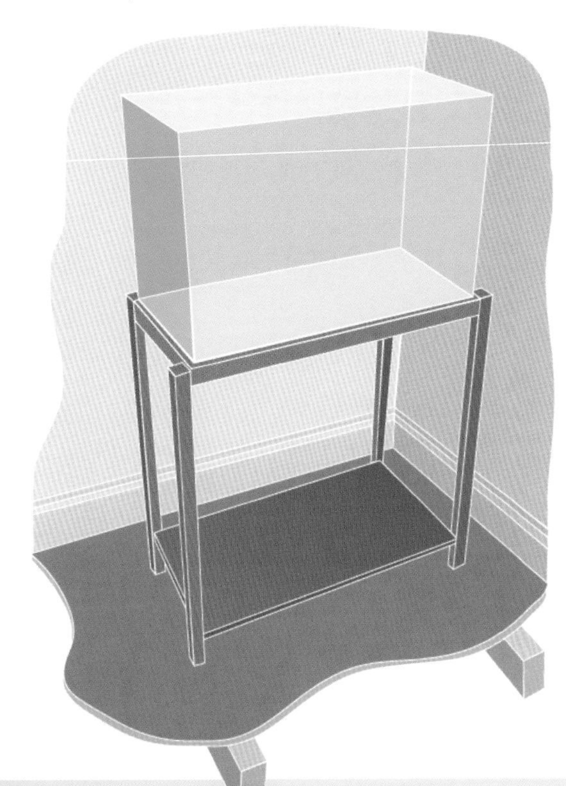

Tip 22 *The built-in frame supports the base of this tank.*

24 Only use small aquariums on upper floors

Be wary of siting anything other than the smallest tank upstairs—the ceiling may not take its weight. The safest positions are adjacent to load-bearing walls, rather than toward the center of the room. If in doubt, consult a structural engineer—the consequences of collapse do not bear thinking about. Also check that your home insurance covers damage from water spillage.

25 Good and bad locations for a home aquarium

Because of the risk of water contamination from cooking fumes, a kitchen is not a good place for an aquarium. Nor is a room with many windows. Here, there will be too much light, which encourages algal growth, and day/night temperatures may fluctuate wildly. A draft-free corner that is shielded from too much direct sunlight in a room you frequent a lot is a far better choice for a home aquarium.

26 Work out the best height for viewing and access

The height at which your aquarium is sited depends on several factors. The tank should ideally be at or just above eye level, so consider from which position, seated or standing for example, you will be doing most of your viewing. Your own height and arm length should also be considered—you need to be able to reach down into the tank for maintenance without standing on a chair or steps, which would constitute a safety hazard.

27 A corner is the best place to site an aquarium

Siting aquariums in the corner of a room not only means they are close to a power outlet, eliminating trailing cables to trip over, but the fish prefer things that way, as they feel more secure without human traffic passing from all sides. Very shy species may benefit further from having the side panels, as well as the back glass, blacked out or at least shielded by tank decor.

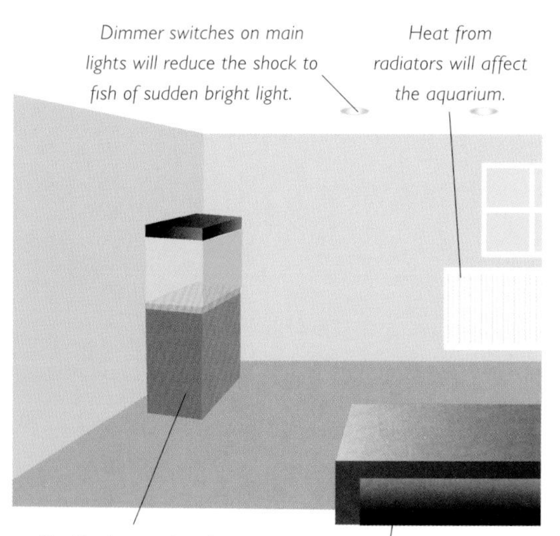

Tip 27 *Siting the aquarium.*

Tip 28 *Themed backgrounds can look gaudy—use them with care.*

Flexible plastic background with black and shaded blue sides.

28 Be sure to fit the background first

A plain, dark background, taped to the external back glass of the aquarium, looks more effective than some of the contrived plastic backgrounds sold off the roll. In either instance, secure the background in place before you move the tank into position against the wall and fill it, otherwise the aquarium will be too heavy to move, and you will have too little space to work.

29 Site the tank as near as possible to a water supply

If you have a choice of suitable sites for your aquarium, always opt for the one nearest your water supply. However carefully you conduct partial water changes, there will always be the risk of spillage, so the smaller the distance between tank and tap, the less chance there is of slopping water onto the floor.

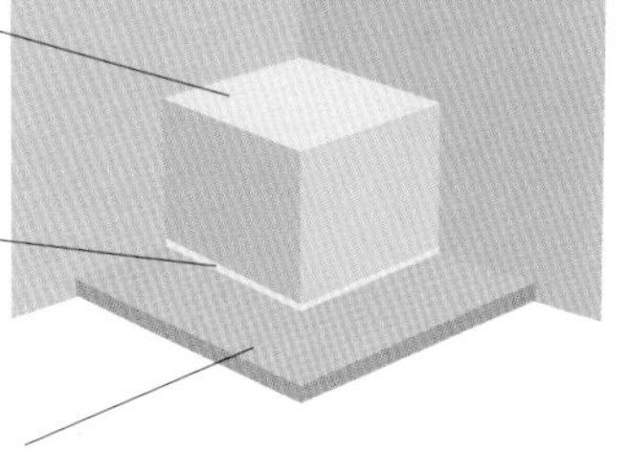

Tip 31 *Tank on a baseboard.*

30 How can I set up an aquarium in a room divider?

Aquariums housed in alcoves or room dividers look attractive, but there is often little or no room to gain access for cleaning and maintenance. If the hood cannot be fully raised, the site is unsuitable. To disguise the necessary space above the tank, you can hook or slot a suitably painted exterior-grade plywood board into place, adopting a similar system to that used in some aquatic stores.

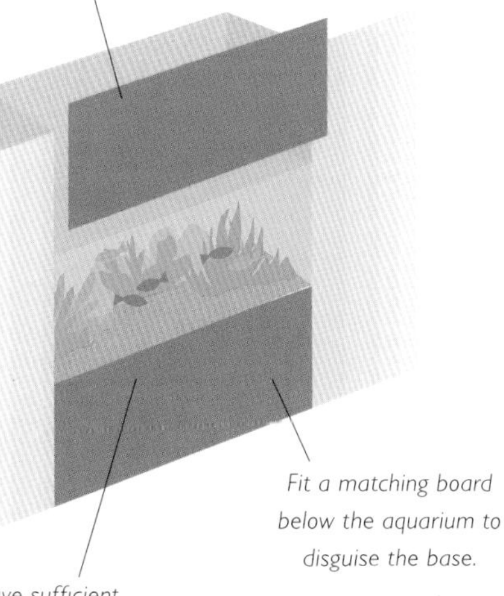

Tip 30 *Making an alcove panel.*

31 Create a buffer zone around a floor-level tank

If aquariums have to be sited at floor level, the glass is at risk of damage (from children's toys being pushed against it, the vacuum cleaner, etc.). It will pay to install the tank on a 2-inch (5 cm) thick baseboard that projects the same distance around all exposed sides to provide a buffer against knocks and bumps. The Styrofoam cushioning can still be cut to the exact size of the tank base.

32 Can I make my own support for an aquarium?

The tank itself is the least expensive component of a system comprising tank, cabinet stand and hood. If you cannot afford a cabinet for a large tank, a sturdy homemade support can be built from dry-laid pillars of bricks or concrete blocks with a 1-inch (2.5 cm) thick baseboard of exterior-grade plywood cut to size. Such supports add substantially to the total weight, and should be used only in ground-floor locations.

Tip 32 *Supporting a tank on concrete blocks.*

33 Matching your aquarium setup to your room decor

Modern aquariums are described as "living pieces of furniture," so bear this in mind when choosing a cabinet. Will it match the rest of your room decor? Cabinets are supplied in light, medium and dark wood effect, in gloss, matt or metallic paint finishes, or even in brushed aluminum, so finding a suitable one should not be too difficult. If your local store does not have the finish you want on display, they can probably order it for you.

Tip 33 *This cabinet is a fine piece of furniture too.*

34 Be safety conscious when setting up your aquarium

Make sure your aquariums are stable and cannot be tipped over by children or pets. Electricity and water are a dangerous mix. Ensure all equipment is plugged into an adapter strip with a surge protector (a power bar with a circuit breaker).

35 Do not store chemical products in the cabinet

With cabinet aquariums there is a great temptation to keep medications and water treatments along with other fishkeeping bits and pieces in the cupboard or shelf space provided. If there are young children or pets in your home don't! Treat medications as you would any other potentially toxic product, and store them out of reach of small fingers under lock and key.

Tip 36
A ready-to-assemble aquarium cabinet.

36 Is it easy to put together a ready-to-assemble cabinet?

An aquarium cabinet is like any other piece of ready-to-assemble furniture—instructions are not always that clear and vital components may be missing. Most cabinets are held together by steel pegs locking onto cams inside the panels. First, lay out all the components on a sheet of newspaper and check that they are all present. Don't rush, and try to enlist help when assembling side and back panels onto the baseboard, as this makes the job much easier. Once the cabinet doors are in place, check that they open and close smoothly before positioning the tank, but be prepared to fine-tune the hinges again when the tank is filled and pressure is bearing down on the cabinet.

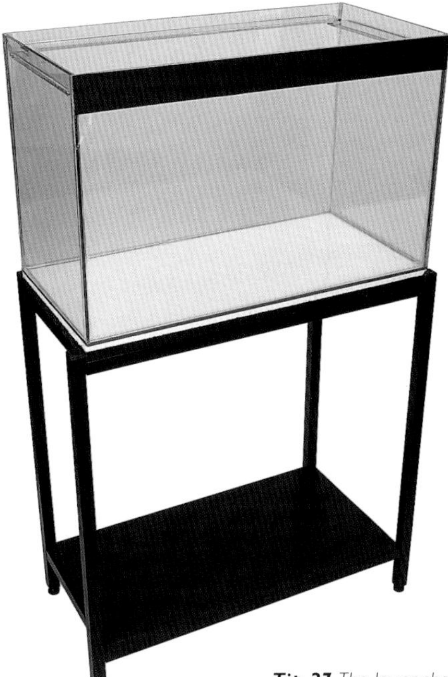

Tip 37 *The lower shelf can take another tank.*

37 Can I stand a tank on the lower shelf of my stand?

Metal box section stands give the option of a second tank on the lower shelf, just above floor level. This can provide space for rearing fry, a refuge for fish that are being bullied or temporary accommodation if equipment fails in the main tank. But be wary of doubling up if you have young children, as the lower tank will be well within their reach.

38 Cutting holes for access in the aquarium hood

If you opt for an external power filter, the aquarium hood will have to accommodate inflow and outflow pipework. Some recesses may already be cut for this, but are they sufficient? If you have to enlarge them or cut extra ones, the structural integrity and appearance of an aluminum hood may be compromised; drilling holes through the back of a wooden or laminate hood is far easier and the results will be neater.

39 Make sure the hood has good access for feeding

Easy tank access encourages good fish husbandry, and it is worth paying a little extra for a hood that allows this. For example, avoid hoods with a feeding hatch over a plastic cover "glass" that cannot be slid aside and will need an aperture cut into it.

40 Can I replace an aluminum hood at a later date?

Hoods made from wood veneer or laminate that sit over the top of the tank are available in most standard sizes. This makes it easy to replace a flimsy aluminum hood with something more substantial when your budget allows. Remember, though, that put-over hoods reduce the visible area of the front tank glass and they conceal the top water level in the same way the tank trim does.

Tip 40 *You can buy replacement tank hoods.*

Two sheets of glass slide in plastic channeling.

Glass marble is siliconed in place as a handle.

Slide back one panel to feed the fish.

Tip 41 *Fitting sliding cover glasses to the aquarium.*

41 Why do I need a cover glass on my aquarium?

Cover glasses perform several functions—they cut down on evaporation loss, prevent escape of fish with wanderlust and act as splash guards for light tubes and fittings. On smaller aquariums, these "glasses" are sometimes made of plastic. Consider replacing them with real glass, which is easier to wipe clean, heavier and less prone to scratching. Have the glass cut professionally, with ground beveled edges to prevent injury. (The redundant plastic condensation tray, shortened to tank height, makes an excellent emergency divider—exactly the right width!)

42 Seek expert help before drilling holes in a tank

If you own large fish with messy feeding habits, sump filtration (located beneath the tank) is an effective way of keeping the water clean. But for this, the tank will need to be drilled to accommodate outlet pipework. Do not attempt this yourself! Ask your aquatic store to do it for you or recommend a specialist with the experience and necessary equipment to do the job properly.

43 Attach a handle to the hood to improve your grip

The lids of aluminum hoods can be difficult to lift for feeding and maintenance, as the handles or recesses provide insufficient grip, particularly for wet hands. A lid slipping from your grasp and banging down noisily will upset your fish. A glass marble or a small, round wooden cupboard doorknob, silicone-sealed onto the lid, is the solution.

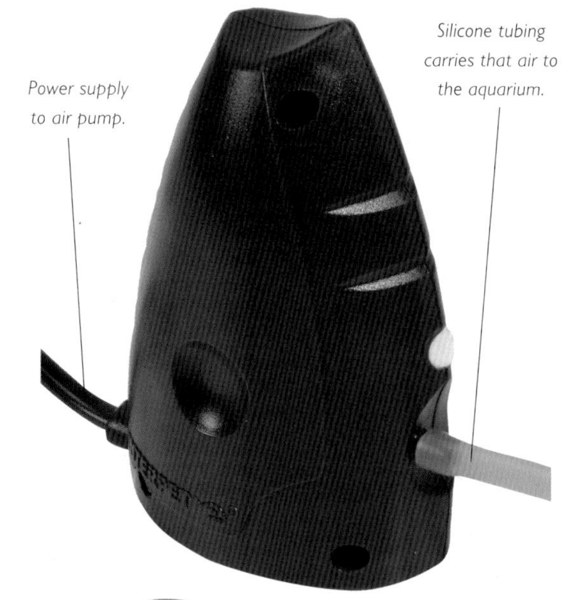

Ideally, mount the air pump above the tank. If not possible, fit a check valve as another way to prevent back-siphoning.

A raised lip on the shelf will stop a vibrating air pump from falling off.

Tip 44 *While the model pictured above is atypical, a standard air pump should be located in the same manner.*

44 What is the best location for air pumps?

Place air pumps above water level (to prevent back-siphoning if the power fails). Never rest them on the hood, where their vibrations will upset the fish (and you, as any noise will be amplified). The best place for air pumps is hanging on a secure hook in the wall behind the tank or on a shelf with raised edges, so that they cannot fall off as they vibrate.

45 How can I protect my filter during a power failure?

During power failures, tanks lose heat very slowly and can be temporarily insulated. More critical is the loss of the filter pump, as beneficial bacteria can start to die off within hours. A U.P.S. (uninterrupted power supply), a piece of equipment designed to safeguard unsaved computer data if the power goes off, will cut in automatically and offers limited power backup to your aquarium. (See also Tip 318.)

Be prepared for emergency tank repairs

A roll of strong, cloth-based adhesive tape is useful for temporarily staunching minor leaks, but you cannot make a permanent repair from the outside with silicone sealant—the tank will need to be emptied, cleaned and thoroughly dried. Then push a bead of aquarium-grade sealant the entire length of the suspect panel. Do not use domestic bathroom sealants as they contain a fungicide that will kill fish.

47 A dehumidifier will control condensation

Open-top tanks, which allow plants to grow above the water surface, are prone to rapid evaporation, and this can lead to condensation in the room that houses them. It is worth investing in a small domestic dehumidifier to remove excess moisture from the air. Never use the condensate water to top up your tank, as it is certain to contain atmospheric pollutants.

A thriving aquarium showing little of the equipment that runs it.

Better Heating and Lighting

48 How do I work out what size heater I need to buy?

To work out the required heater wattage for an aquarium, use an approximate guideline of 4 watts per gallon (1 watt per liter) of tank volume. Round up from this figure to the nearest standard heater size. This guideline will be sufficient for most situations; however, if the ambient room temperature is on the cool side, a higher wattage may be required.

49 How many heaters should I use in my aquarium?

For larger aquariums, consider splitting the total required heater wattage between two smaller heaters. Apart from providing more even heating, this will prevent the temperature from changing too quickly in the event of a heater failing. If one of the heaters does fail, the tank temperature will not change as quickly since the other heater will be operating normally, allowing you time to notice and cure the problem.

50 Do I need a guard on my heater?

It is sensible to use a heater guard especially if you have large, boisterous fish that might accidentally break your heater. A guard will also help to prevent damage to the heater during tank maintenance or when moving heavy decor around. Some fish may also rest against exposed heaters and burn themselves without being aware of it. (See also Tip 207.)

Tip 53 *A fine heater cable can be buried beneath the substrate.*

51 Monitor the temperature of the tank every day

Use a thermometer to check the tank temperature on a daily basis. Even if the heater has an accurate temperature setting dial (most heaters are sold with built-in thermostats, often called heaterstats), the actual temperature achieved in the tank will vary depending on the ambient temperature, circulation efficiency, etc. Position the thermometer so you automatically see it when you pass the tank. Any major change in the thermometer reading will help you to spot a potential heater failure before it affects fish and plants.

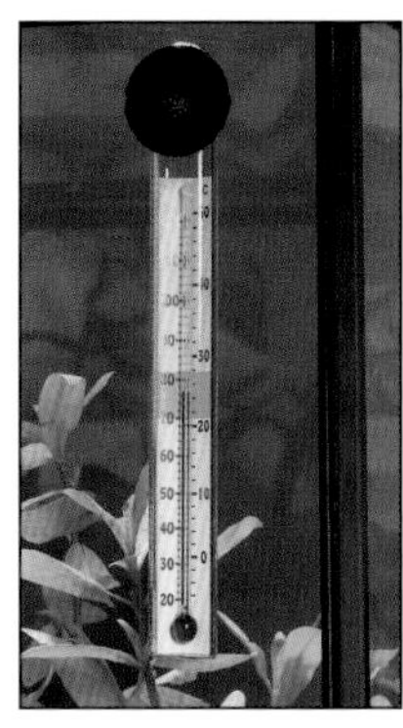

Tip 51 *Install the thermometer where it is easy to see.*

52 Does the heater have to be inside the display aquarium?

External heaters that can be connected into filter return lines are available, as are external canister filters (called thermofilters) that incorporate built-in heaters. Using one of these reduces the amount of equipment visible inside the tank, making for a more natural display. Having the heater outside the main tank also reduces the risk of heater breakage. (See also Tip 86.)

53 Underfloor heating for aquariums

Although not always easy to find, an additional heating method may be incorporated in planted tanks to keep the substrate layer warmer. This involves the use of a low-wattage heating cable placed beneath the substrate. It is usually set a degree or two above the main heater to encourage warm convection currents to rise through the tank. These slow currents can help to keep the substrate healthy and boost plant growth. (See also Tip 135.)

54 Calibrate your aquarium thermometer for accuracy

Aquarium thermometers are often inaccurate by plus or minus a couple of degrees. If you have several tanks or temperature-sensitive fish, it is worth investing in a scientific-grade mercury thermometer and using it to check the accuracy of the aquarium thermometer. You can label the tank with the amount the aquarium thermometer is out.

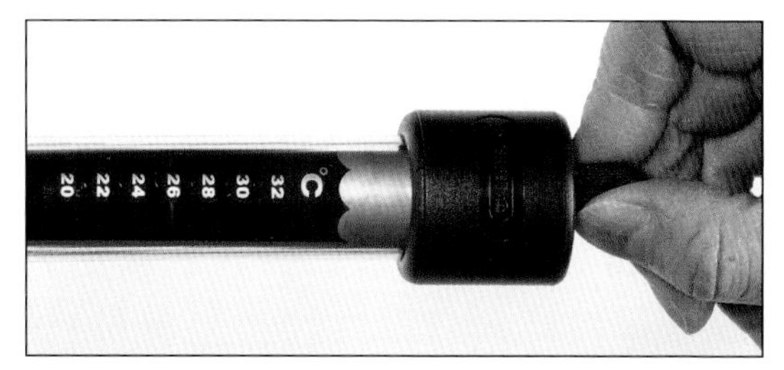

Tip 54 *Set heater temperature and calibrate thermometers.*

55 Use holders to keep heaters away from the glass

Always use heater holders to make sure your heaters are not in contact with the substrate, rocks or the aquarium glass. If the heater tube is in contact with anything solid (or buried) a hot spot will develop and the tube may crack so the heater burns out, or worse, the water becomes live with electricity. And a heater touching a tank pane can crack the glass!

56 My heater turns off below the set temperature. Why?

If the thermostat on a heaterstat switches off at a temperature lower than set, it could indicate a fault. More likely, though, the heater is located where water circulation is sluggish, so it raises a small pocket of water to the set temperature and then turns off. The solution is either to move the heater or adjust the water return from powerheads or filters to alleviate the dead spot in the water circulation.

57 Allow heaters to cool before removing them

Always remember to switch off the heaterstat about 15 minutes before you remove it from the aquarium for any reason to allow it to cool. Heaters removed suddenly from the water while still hot could crack. Some newer designs have safety cutouts that should prevent this type of breakage, but it's always better to be safe than sorry!

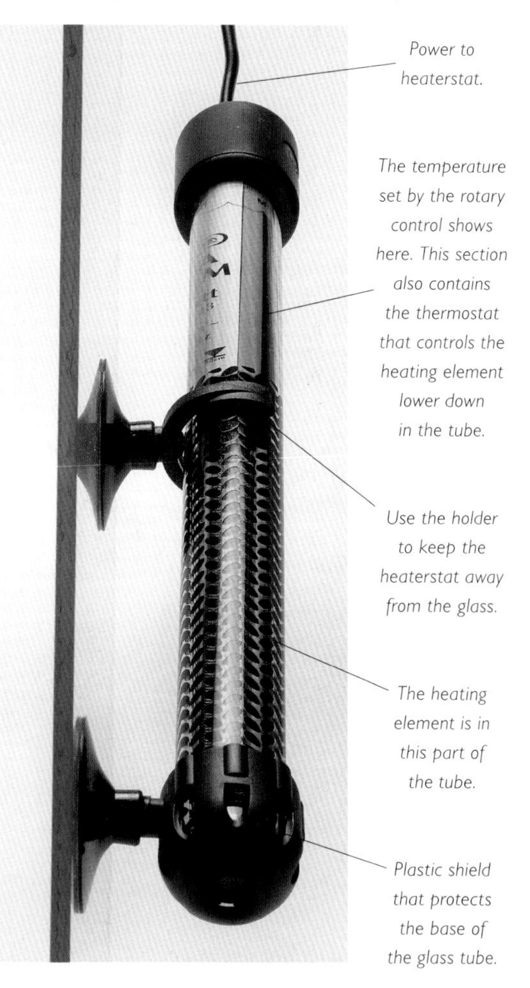

Tip 55 *Use holders to keep heaters away from glass surfaces.*

Better Heating and Lighting

58 Fluorescent bulbs—ideal for aquarium lighting

Fluorescent bulbs are the best option for aquarium lighting in the majority of situations, and they are by far the most widely available. Some bulbs provide a more suitable light spectrum for photosynthesis and plant growth than others. These are usually bright "daylight" with peaks in the red area of the light spectrum. Bulbs of the wrong spectrum may encourage algae growth rather than good plant growth.

59 Combine fluorescent bulbs for an all-around lighting effect

Combinations of different fluorescent bulbs can provide the most effective all around lighting for the tank. Using a bright white bulb in combination with a less-bright warmer colored bulb will bring out the best colors in fish, as well as provide a good overall spectrum for plant growth.

Tip 59 *White and pink bulbs combined.*

60 Lighting levels in tanks containing plants

For tanks containing live plants, the required wattage for lighting can be at least double that required for non-planted tanks, depending on the type of plants. So allow up to 20 watts per 1 square foot (900 cm^2) of tank surface area, or about 3 watts per 1 gallon (4 L) of tank volume. These guidelines apply to tanks up to 18 inches (45 cm) deep. Appropriate fertilization and carbon dioxide levels are necessary to allow aquarium plants to make full use of very bright lighting. (See also Tip 138.)

Tip 60 *Adequate lighting is essential for healthy plant growth.*

How plants absorb white light spectrum

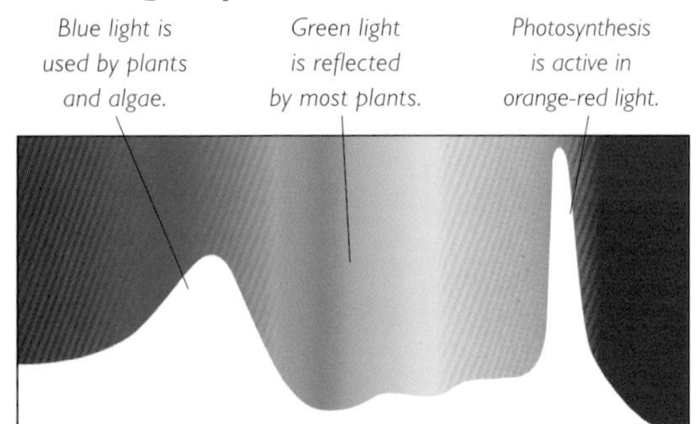

Colors in white light spectrum.

Spectrum produced by a white triphosphor fluorescent bulb

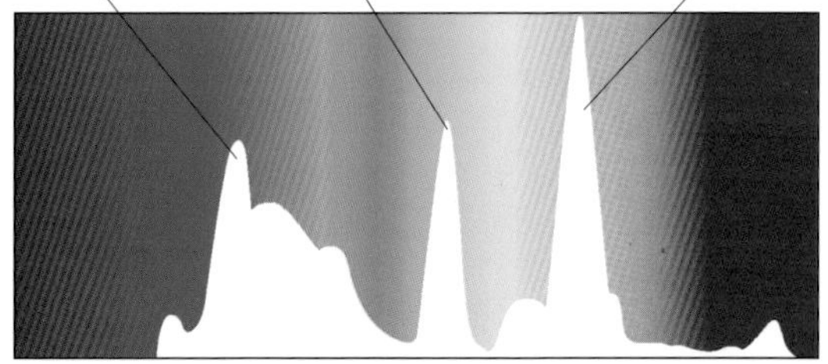

Colors in white light spectrum.

61 Lighting levels in tanks without plants

For tanks that do not contain live plants, the aquarium lighting need only be sufficient to see the fish and to give a natural day–night cycle. About 10 watts of fluorescent lighting per 1 square foot (900 cm^2) of tank surface area or about 1.5 watts per 1 gallon (4 L) of tank volume is sufficient. In the absence of live plants, brighter lights will just encourage algae growth.

62 Do I need a timer to operate aquarium lighting?

Timer units are strongly recommended for aquarium lighting. They will ensure a consistent lighting period for your fish and plants every day without the need to remember to switch the lights on and off manually.

63 Provide a midday peak for planted aquariums

For well-planted tanks with multiple bulbs, an effective strategy is to stagger the lighting intensity with timers. The lighting can then be built up gradually to a peak in the middle of the day and then reduced to give a more natural light cycle. A single bulb can be used for the first and last hour, and additional bulbs turned on for the brightest part of the day.

64 How long should the lights be left on each day?

In an aquarium containing live plants, the lights should be on for 10–12 hours to provide a natural day and night cycle. Appropriate dark and light periods are essential for good plant growth. In fish-only tanks the lighting period is not as critical, as long as the fish have some sort of normal day–night period. As long as the aquarium does not receive bright ambient daylight, you can shift the illuminated period toward the later part of the day to provide a good viewing session during the evening.

Tip 64 *The typical tropical day is lit for about 12 hours from 7 a.m. to 7 p.m.*

Shifting that period to 11 a.m. to 11 p.m. provides evening viewing.

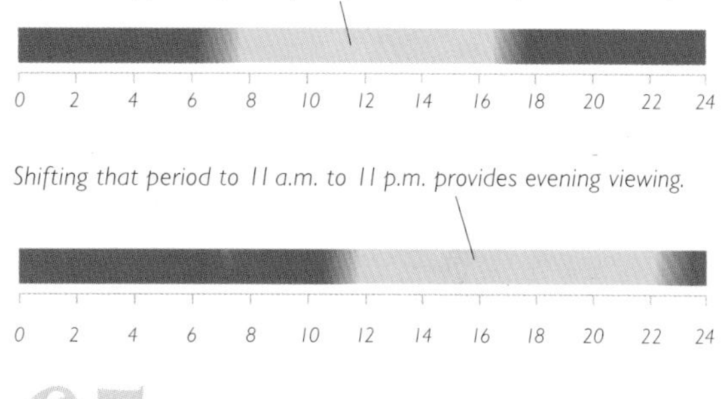

65 Provide dim aquarium lighting for nocturnal fish

Some aquarium fish prefer dim lighting in the tank and may be nervous under bright lighting. These include many of the catfish family and a number of other groups of fish that are normally nocturnal. For these fish it is better to keep the tank lighting to a minimum (normal daylight is often sufficient). Planted tanks can incorporate floating plants to provide shade.

Tip 65 *Bristlenose catfish* (Ancistrus temminckii) *will thrive in subdued aquarium lighting.*

Better Heating and Lighting

66 Achieve different lighting effects in the aquarium

For evening viewing, or to encourage shy or nocturnal fish to venture out, blue bulbs (also called actinic or moonlight bulbs) can be used to great effect. These bulbs are also useful in combination with a normal, brighter bulb to provide a dusk and dawn effect on either side of the main lighting period to give a more natural transition between light and dark. This reduces the chances of fish being startled by sudden changes in light.

67 Reflectors make the most of the available light

The use of reflectors behind fluorescent bulbs can increase the overall light entering the tank by directing as much of the light energy as possible down into the tank. Curved reflectors for mounting above lights are supplied by several manufacturers of aquarium lighting for this purpose. They are made from aluminum or flexible plastic and have an optimum shape to direct the maximum amount of light down into the aquarium.

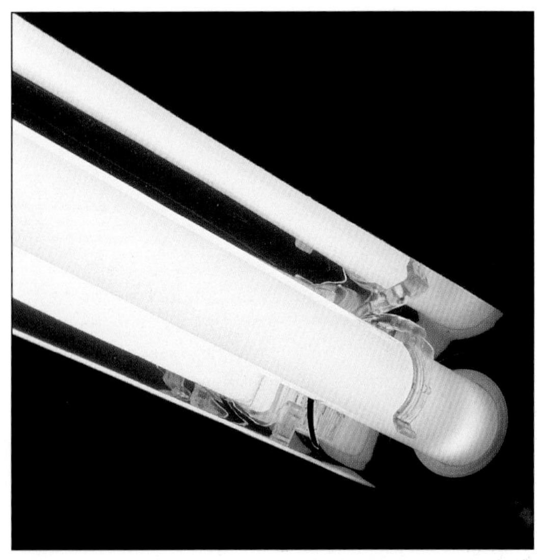

Tip 67 *Curved reflectors maximize light from bulbs.*

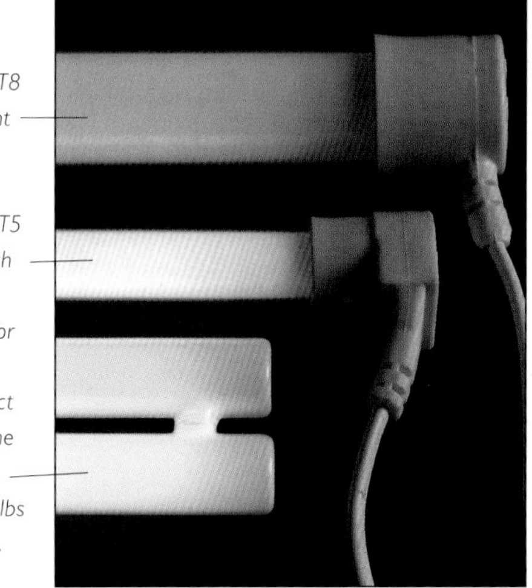

Tip 68 *Narrower bulbs provide brighter light.*

68 Use narrower, more efficient bulbs for planted tanks

Standard fluorescent bulbs, known as T8 size, are 1 inch (25 mm) in diameter. However, narrower T5 bulbs at $^5/_8$ inch (16 mm) in diameter are now available and provide a much higher output for their size. They are especially useful for heavily planted tanks that require bright lighting. An even more compact form of these bulbs, known as "biax" or "PL," is essentially a linear T5 tube bent back on itself with four pins at one end. These can combine two colors in one unit.

69 Deep tanks need brighter light

Deeper tanks require more intense lighting. In general, tanks 24 inches (60 cm) deep or more will need higher wattages, and the use of reflectors is highly recommended. The clarity of the water also has more effect on light penetration in deeper tanks, so tannin-stained water, for example, could reduce light penetration significantly. T5 linear or compact PL fluorescent bulbs are very effective for these deeper tanks.

70 Replace fluorescent bulbs regularly for good lighting

The light intensity and spectral output of fluorescent bulbs deteriorate over time. This will affect plant growth before the reduction in output becomes apparent to the human eye. You should therefore replace bulbs in planted tanks every 6–12 months to maintain full lighting spectrum and intensity. (See also Routine Maintenance, page 115.)

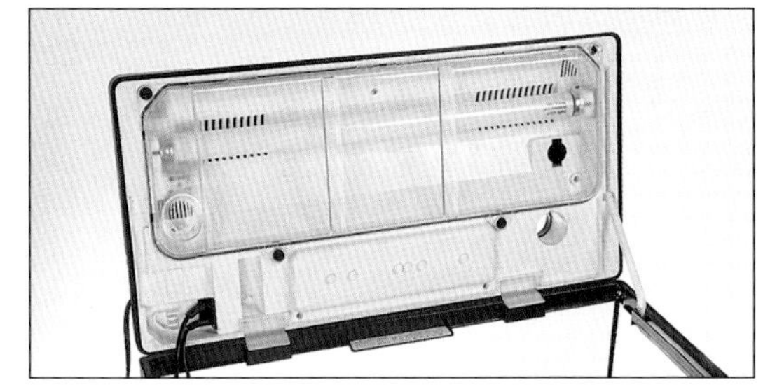

Tip 71 *Lights within a waterproof compartment.*

71 Keep condensation covers clean for maximum light

If the tank has plastic or glass condensation covers between the lighting and the tank, keep these clean to maximize light entering the tank. Dirty covers greatly reduce the amount of light reaching the tank. Some lighting systems reduce the need for covers by incorporating fittings within a transparent waterproof compartment.

72 An alternative to pendant lighting in a confined space

If you want the plant-growing benefits of an open-top tank without having to install pendant lighting, an over-tank luminaire is the answer. This is a fully self-contained hood, complete with built-in reflector and splash glass, that is mounted on slim, raised brackets at either end of the tank so it can be tilted back for access. Switches and ballast are built into the unit, which can accommodate two or four linear bulbs.

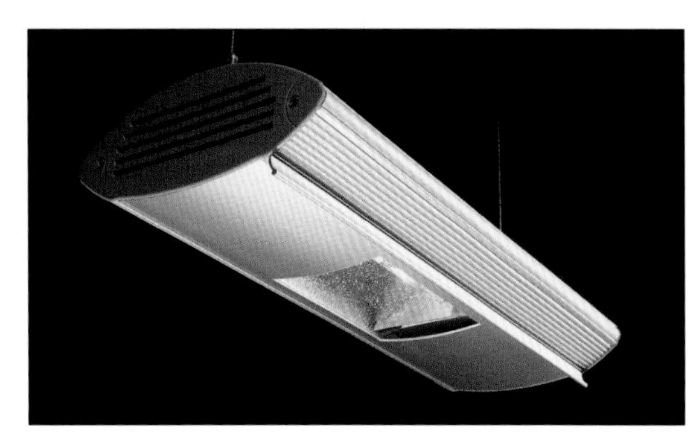

Tip 73 *Metal-halide lighting is ideal for deep, planted tanks.*

73 Metal-halide lighting—costly but effective

Metal-halide lighting can provide the ultimate in bright lighting for planted tanks, and is especially suited to deep tanks. This type of lighting is normally suspended above an open-top aquarium (partly because of the heat generated). Metal-halide lighting systems are more expensive to buy initially, and the higher wattages make them more costly to run, but they do provide excellent lighting.

Tip 72 *An over-tank luminaire.*

Better Water Management

74 Buy the right size and type of filter for your aquarium

The filtration system is the essential life support of an aquarium, so it is important to match the size and type of filter to the size and type of tank in which it will be used, as well as to the number and species of fish kept. No single filter type is perfect for every setup. It is a good idea to opt for a slightly larger filter than recommended for your tank size to ensure adequate filtration.

75 What does a filter do to keep the water clean?

A filter can potentially perform three basic types of filtration which are usually referred to as mechanical, biological and chemical. Mechanical filtration removes physical dirt particles from the water and helps to keep the aquarium looking clean. Biological filtration is the most important function, as it converts invisible toxic wastes, such as ammonia, to less toxic forms, such as nitrates, through the action of beneficial bacteria that build up within the filter material. Chemical filtration can be used to remove specific substances from the water.

76 Maturing a new biological filter the natural way

No matter what type of filter system you use, it takes time for a new biological filter to build up a stable population of beneficial bacteria—typically four to six weeks or even more. It is therefore essential to stock the aquarium with fish gradually and allow the bacterial population to increase steadily to cope with their waste. The maturing time can be shortened by adding a small amount of medium from a mature filter, or by squeezing out the "dirt" from a mature sponge onto a new one. This seeds the new filter with essential bacteria.

77 Bacteria cultures can help but are not always essential

Filter bacteria starter cultures are popular with those too impatient to wait at least four weeks for their tank to mature naturally (and for equipment teething troubles to be sorted). Some cultures are recommended as a means of replacing lost bacteria after water changes or filter maintenance. But even if you "throw away" half your bacteria, the remainder can divide and double their population in a few hours.

Tip 75 *Efficient filtration will keep your aquarium looking clean.*

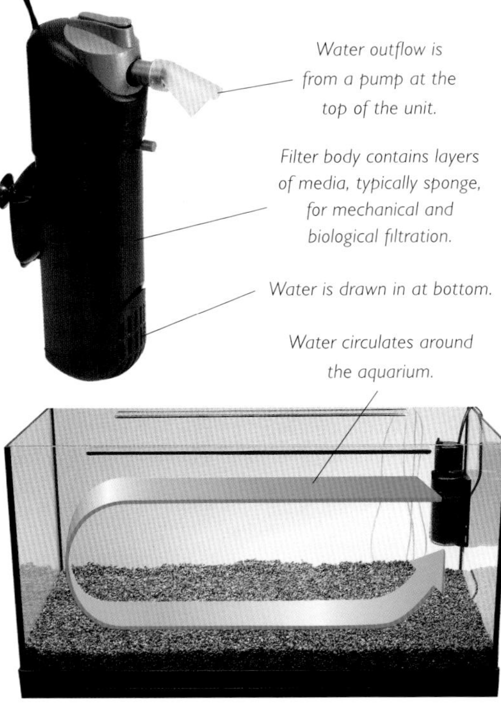

Tip 78 *An internal power filter in the rear corner works well.*

78 What is the best type of filter for a small aquarium?

Internal power filters are ideal for small- to medium-sized aquariums. These filters provide good circulation, as well as mechanical and biological filtration, usually with sponge media. Some designs also allow for the inclusion of additional filter media for specific purposes, such as activated carbon.

79 Do I really need an external filter for my aquarium?

For larger aquariums, and those requiring plenty of biological filtration (e.g. those with large and messy fish), external canister filters are usually the filter of choice. They have a generous capacity for biofiltration and plenty of room for the media to be tailored to suit the tank setup.

80 Should I arrange the filter media in any particular order?

In any multistage filter (such as canister filters), place the mechanical media before the biological media in the water flow. In this way the mechanical media remove larger particles first, and prevent rapid clogging of the biomedia with dirt, which would reduce its "bio" efficiency. In smaller filters, mechanical and biological filtration are often combined, such as in sponge media.

81 Can I combine two sorts of filters in my aquarium setup?

In some situations, a combination of filter types may provide the best solution. For example, the combination of internal power filters and external canister filters provides plenty of biological filtration, as well as excellent mechanical filtration and good circulation. Combining filters also provides a backup should one fail for any reason.

82 Cleaning filter media without harming the bacteria

Never clean or replace all your filter media at once, as this will eliminate the useful filter bacteria that break down wastes. If you have two filters clean them alternately, otherwise clean only a third or half of the media at a time. Where a filter is provided with more than one section of biomedia (e.g. twin sponges), clean them alternately. Rinse dirty media in old tank water during a water change—it will be the right temperature and not contain bactericidal chlorine. (See also Tip 439.)

83 How does a fluidized sand bed filter work?

Fluidized sand bed filters provide excellent biological filtration, using a constantly moving bed of fine sand particles that become colonized by bacteria. The very high surface area and constant motion of the particles make it a highly efficient biofilter. This type of filter, also used for ponds and saltwater tanks, requires mechanical prefiltration to work efficiently.

Tip 83 *A fluidized sand bed filter.*

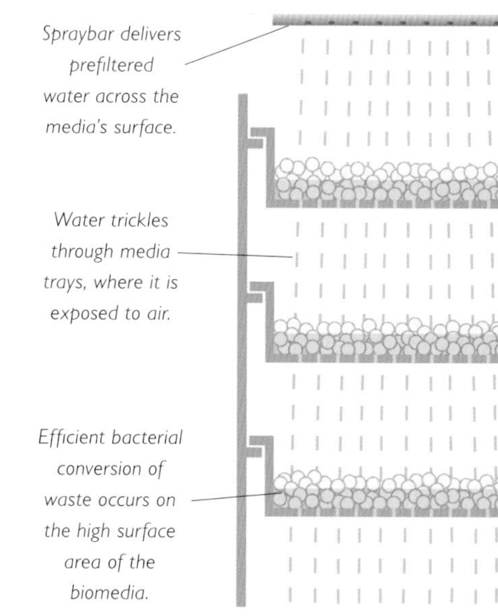

Tip 84 *How a trickle filter works.*

84 I have heard of a trickle filter, but how does it work?

Trickle filters—and wet/dry filters—are efficient biofilters that rely on the principle of exposing the biomedia to oxygen in the air to boost the rate at which oxygen-loving bacteria break down waste. Various designs are available and, as with fluidized sand beds, the water needs to be mechanically prefiltered for best results.

85 24-hour filtration is essential to keep bacteria alive

Filter bacteria require oxygen 24 hours a day. If you switch off your air pump or filter at night (for example, because of the noise), your filter bacteria will die—they can live for only a couple of hours or so without water flow bringing them oxygen. So if your tank is in your bedroom, use a silent filter and hang the air pump out on the landing!

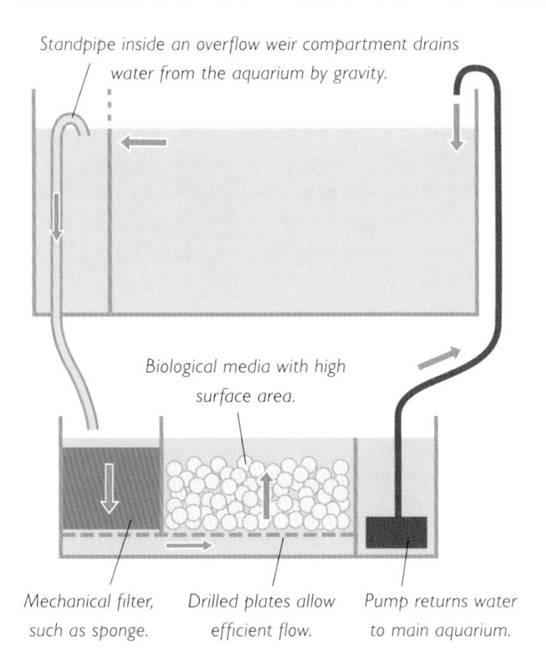

Tip 86 *An aquarium with a sump filtration system.*

86 Sump filters offer the ultimate filtration system

For larger tanks, a sump filter may provide the best filtration option. This is essentially a second tank placed beneath the main aquarium. Water drains from the main tank by gravity and is returned via a pump in the sump. A very large amount of media of different types, plus equipment such as heaters, can be placed in the sump, allowing an uncluttered display aquarium.

87 Air-powered sponge filters are ideal for fry-raising tanks

Air-powered sponge filters are an excellent choice for small fry-raising tanks—unlike power filters, there is no danger of tiny fry being sucked in. These filters will provide some mechanical filtration and circulation, as well as biofiltration, once the sponge is mature. (See also Better Breeding.)

88 Air-powered box filters offer versatile, small-scale filtration

Another useful air-powered filter is the inexpensive box filter. These are also powered by an air pump and allow you to include whatever media you need for a particular purpose. Media could include activated carbon or adsorptive resins for chemical filtration, filter wool for mechanical filtration and other media for biological filtration.

Tip 88 *A typical air-powered box filter.*

89 What is undergravel filtration? How does it work?

Undergravel filtration (U.G.) works by drawing water down through the gravel bed, which acts as the filter medium, and through an uplift tube that uses either a column of rising bubbles created by an air pump or (for more efficient water flow) direct suction from a powerhead (a water pump). In all but the smallest tanks at least two uplifts should be used (one in each rear corner) to maintain an even flow through the gravel bed. U.G. filtration was once the mainstay of aquarium filtration, but there is now a wider range of filtration options available, which have advantages over it in some situations.

90 Do I need a special substrate for U.G. filtration?

If U.G. filtration is used, the substrate must be composed of standard gravel with a particle size of about $^1/_{12}$–$^1/_8$ inch (2–3 mm) to allow efficient water flow through the bed. Finer gravels will tend to clog more quickly and larger grade gravel allows debris to fall between the particles too readily. Fine sand cannot be used as an aquarium substrate where U.G. filtration is employed, as the fine grains will fall through the U.G. plates that hold the substrate bed clear of the tank base.

91 How to prevent digging fish undermining U.G. filtration

If U.G. filtration is used in tanks containing fish that like to dig, such as many cichlids, it is a good idea to include a gravel tidy between two layers of substrate. This is simply a sheet of flexible perforated material that will prevent the gravel bed from being completely excavated by digging fish, which would "short circuit" the filter bed. (See also Tip 185.)

92 The benefits of reverse-flow U.G. filtration in the aquarium

A disadvantage of normal U.G. filtration is that dirt is sucked down into the gravel bed. Reverse-flow U.G. filtration overcomes this problem because water is mechanically prefiltered in a power filter and then pumped down the uplift tubes and up through the gravel bed. This system provides good mechanical and biological filtration, but surface agitation is limited, so additional aeration is needed. (See also Tip 155.)

93 Aquarium decor can function as a filter medium

The aerobic bacteria that process waste in the filter also colonize oxygenated surfaces in the aquarium. Hence it is healthier to have gravel (which offers a greater colonizable surface area compared to glass, even without undergravel filters) and decor than a bare tank. And during routine maintenance just vacuum mulm from the gravel surface—don't churn the gravel and bury the surface bacteria as they will die from lack of oxygen.

All the surfaces in the aquarium, including bogwood, rocks and gravel, will become colonized with beneficial bacteria.

Tip 93 *Even the decor acts as a biological filter.*

94 Provide water circulation to suit your fish

The circulation in an aquarium should be appropriate to the type of fish inhabiting the tank. Some fish come from still or slow-moving water, and may become stressed if they are constantly swimming against strong currents in the tank. Conversely, some fish originate from fast-flowing waters and appreciate good circulation and highly oxygenated water.

Tip 94 *Arulius barbs* (Barbus arulius) *enjoy water currents in the aquarium.*

95 Is the turnover rate of filters important?

Filters are often rated on their pump circulation rate, but this is not the only important factor. The volume and type of media are also important, especially for efficient biofiltration. For any given volume of water (any size tank) the recommended filter turnover rate is usually lower where large external canister filters are employed than when smaller internal filters are chosen. This is because the larger media capacity and longer contact time in external canisters allow more waste conversion with each pass through the filter.

96 Do I need to agitate the water to boost oxygen levels?

It is not necessary to have vast amounts of aeration and turbulence to ensure the water contains enough oxygen. The important points are good, even circulation and movement of the surface—this increases the surface area available for gas exchange. The normal return of water from the filter along the surface or via a spraybar will be adequate, even for fish from high-oxygen environments.

97 When would I need to supply extra aeration?

Extra aeration is more important during hot summer weather, because water holds less oxygen as temperature increases. Fish may also benefit from additional aeration when being medicated, as both diseases and methods of treatment may otherwise result in hypoxia (oxygen shortage) in the fish. Directing the filter return along the surface, or using a spraybar or airstones, greatly increases the surface area of the water that's in contact with the air, and thus its oxygen uptake capacity.

How temperature affects oxygen levels

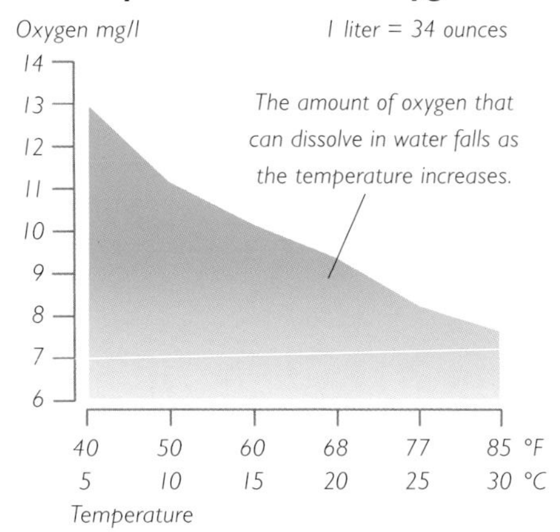

Tip 97 *An internal filter outflow provides surface agitation.*

Better Water Management

98 It is best to avoid strong aeration in a planted tank

Extra surface agitation from airstones, spraybars and other methods that create strong aeration are less suited to planted tanks (at least during the "daylight" hours), as they will increase the loss of carbon dioxide from the water, and plants require this gas for photosynthesis. (See also Tip 138.)

99 Can I fill my aquarium straight from the tap?

Water companies add chlorine and/or chloramine, both toxic to fish, to tap water as a disinfectant to make it safe for us to drink. Both can be removed using a chemical conditioner, but chlorine can also be driven off by running the water hard into a bucket, or by aerating it in a bucket or bin overnight. Note that not all dechlorinators remove chloramine, and all tap water must be treated before use.

Tip 99 *Dilute tap water conditioner before adding it to the tank.*

100 Aquarium decor may affect water chemistry

Many items of aquarium decor can affect water hardness and pH (the degree of acidity or alkalinity). These include some rocks and substrate materials, clay pipes and flowerpots, which usually increase hardness and pH, while wood, cork bark and coconut shells will soften and acidify water. Check any suspect item by leaving it in a bucket of water for a week or two. Test hardness and pH before and after to see if there is any effect. Gravel (and rocks) can be tested for the salts that cause hardness by adding a strong acid to a sample and seeing if it fizzes. Take care when using acids; if needed seek professional help, such as from a chemistry teacher or pharmacist. (See also Tip 195.)

Tip 101 *These Rift Lake cichlids need hard water.*

Not all water is the same as far as fish are concerned

Never forget that not all water is the same, and different fish come from different types of water, from soft and acid to hard and alkaline. Although some are hardy or have been acclimatized to "unnatural" water through generations of captivity, many will do badly or even die if the water conditions are wrong. Always check what water parameters a fish needs before buying.

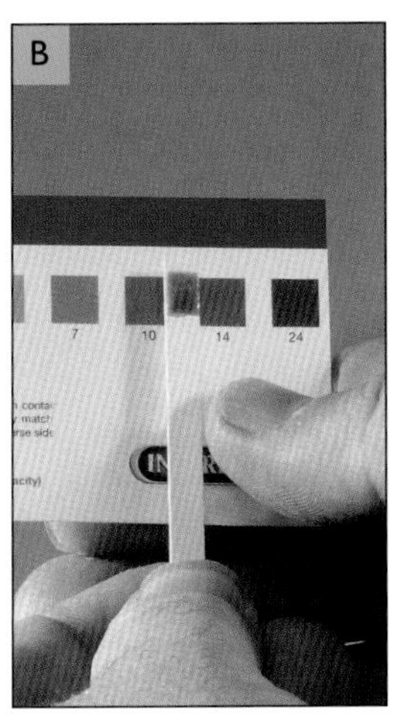

A *Take a fresh test strip from the sealed pack and dip it briefly into the aquarium water.*

B *Compare the color change with the printed chart. This strip shows a water hardness of 14°dH.*

Tip 102 *Taking a dip test for water hardness.*

102 Water hardness scales can be confusing

The value of a degree of general water hardness varies from country to country. For example, a German degree represents almost 20 percent more calcium carbonate than a British degree. So check which degrees your test kit measures, and, if necessary, convert the hardness figure recommended in the literature to the hardness your kit measures.

103 Use the same test kits for consistent results

Nowadays, most test kits are reasonably accurate. However, when you want to make a series of comparative measurements (e.g. weekly nitrate measurements or pH before and after any adjustment), always use the same brand of test kit to ensure that you are measuring to the same level of accuracy.

104 The pH scale is more sensitive than you think

Remember that the pH scale is logarithmic; in other words, every unit change represents a ten-fold increase in acidity or alkalinity. Thus, for example, pH 9 is ten times more alkaline than pH 8 and one hundred times more alkaline than pH 7. Hence, when increasing or decreasing pH with fish present in the tank changes should be made very gradually.

The pH scale runs from 0 (extremely acid) to 14 (extremely alkaline). Neutral is pH 7.

pH 9

pH 8

pH 7

105 Make sure your nitrate test kits aren't expired

Nitrate test kits can be difficult to read and have a relatively short shelf life. Nitrate is produced continually where life (and hence waste) is present, and unless there is a perfect balance between fish and plant life (very unlikely) or nitrate-removing resins are used, a zero nitrate reading is improbable. If your nitrate level is apparently zero, test again and/or buy a new test kit.

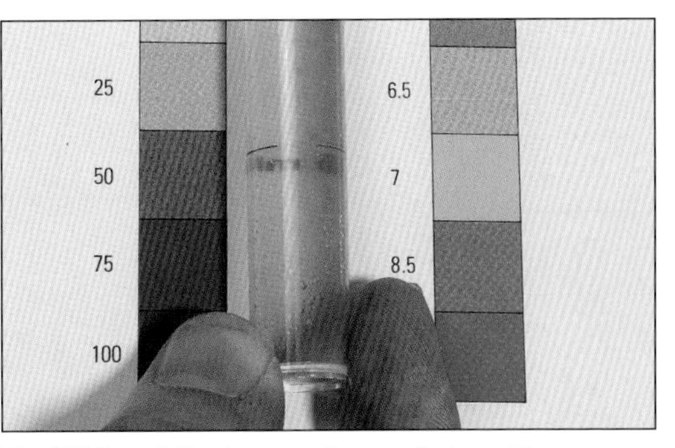

Tip 105 *For reliable nitrate readings use fresh test kits.*

106 Nitrate shock can be a major threat to fish

Your established fish are healthy, the hardness and pH of your tank suit the fish you just bought, and yet the new fish fall ill and maybe die within a couple of days. The answer is normally a long-term buildup of nitrate in the aquarium. The resident fish have adjusted as the level gradually rose, but the new fish experience nitrate shock at the sudden increase.

107 Reverse osmosis water is too pure to use on its own

Reverse osmosis (RO) removes everything from water, including oxygen, and if used neat it may cause fish to suffocate. Always aerate RO water before use. In soft water tanks, water with no dissolved salts will also be unstable with regard to pH, so mix it with some tap water before use or include a little calcium carbonate (e.g. coral sand or dolomite chips) in the filter as a pH buffer. Take regular tests to check the water quality.

Tip 107 *A reverse osmosis unit sieves out any contaminants.*

Tip 108 *Once cleaned, rainwater is ideal for soft-water fish.*

108 Can I use rainwater directly in my aquarium?

Rainwater can be used for keeping soft-water fish in areas where the tap water is hard and alkaline, but always treat it before adding it to the aquarium. Strain it through a fine linen handkerchief into a large, inert receptacle set up with an internal power filter filled with activated carbon, and run this for at least a couple of hours to remove any pollutants.

109 Zeolite is useful as an emergency filter medium

Zeolite is a naturally occurring ion-exchange resin that removes the toxic ammonia continually produced by fish and waste. It should not be used as an alternative to biological filtration, but it can be very useful in emergencies (as a filter medium) or in situations where filtration isn't possible—e.g. fish bags, tiny tanks for hatching and rearing or hospital tanks where bactericidal chemicals are in use.

110 Use resins to remove specific substances

Specific ion-exchange resins can also be used in the filter to adsorb particular substances. Resins are available that adsorb both phosphate and silicate, or both ammonia and nitrate. They may be useful in controlling these substances where they are present in your tap water or other supply. They can be placed in fine mesh bags and should be positioned to allow maximum water flow through them.

Tip 110 *Resins such as this remove nitrates or phosphates.*

111 What does activated carbon do and when should I use it?

Carbon does not remove biological waste (ammonia, nitrite, nitrate) contrary to popular belief. It does remove some heavy metal salts, methylene blue and some other medications. Heavy metals should never be allowed into the aquarium; medications are best used in a hospital tank and never continuously in the main aquarium. Thus, it is pointless to use expensive carbon in your filter except when specifically required.

Tip 111 *Activated carbon from various sources is the most effective.*

112 Can I use domestic water softeners for the aquarium?

Beware of using domestic products based on the ion-exchange principle to soften water for the aquarium. While the resins remove calcium and magnesium, they replace them with sodium ions, which soft-water fish don't relish. A safer method is to use peat moss in a cloth bag, aerating the water in a holding receptacle for at least a week and then filtering the water through activated carbon to remove tannin staining.

Tip 112 *Angelfish* (Pterophyllum scalare) *enjoy soft water.*

113 Can I use any type of peat to soften aquarium water?

You can use horticultural peat moss instead of aquarium peat to acidify water. It is cheaper, just be sure to check (on the packaging or by writing to the manufacturer) that it contains no additives. Use peat in a fine net bag in your filter—fill the bag and then rinse it thoroughly to remove the dust that goes through the mesh. Or boil peat in an enameled or glass pan (metal may react with acid peat) to create "blackwater tonic." (See also Tip 388.)

114 Medications can have negative side effects

Many aquarium medications, including some sold for combating fungus, protozoa and larger parasites, will kill bacteria—including the friendly bacteria responsible for biological filtration. Hence, if these chemicals are used in the main aquarium they may lead to ammonia and nitrite problems, perhaps more serious for the fish than the pathogen being treated. So, if possible and where appropriate, use medications in a hospital tank only.

115 Make a net bag to contain filter media

You may read that peat and other filter media should be put in a fine net bag to keep them where intended. But, where do you get the net bag? You can obtain net bags from your aquarium dealer, but make sure they are fine enough to contain peat. Another strategy is to use a new pair of nylons or thin tights, or part of one, with the end(s) knotted.

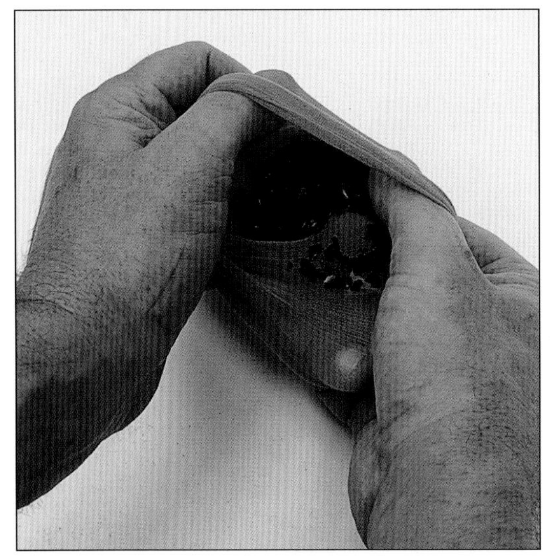

Tip 115 *Using a pair of nylons as a net bag for carbon.*

Tip 118 *Clear water is one sign of a healthy aquarium.*

116 Suspended particles in the water can cause problems

Suspended particles can give rise to skin and gill irritation, causing fish to scratch as if they had parasites. Most scratching is, however, environmental rather than parasitic. Suspended particles of mulm indicate inadequate filtration. Adding coral sand to increase alkalinity will break down into irritant particles so it should be mixed with the gravel (10–15 percent coral sand to 85–90 percent gravel).

117 Change the pH of aquarium water gradually and naturally

Disasters can occur with proprietary chemical pH adjusters. It is safer to use natural materials to adjust and buffer the pH gradually. Peat can be used to acidify the water. Calcium carbonate in the form of coral sand, crushed shell or dolomite chips can be used in a filter or mixed in the substrate to increase or buffer alkalinity for hard-water fish such as Rift Lake cichlids.

118 The water in my tank is cloudy. Should I be worried?

A new aquarium commonly develops cloudy water due to residual dust from the gravel, a bacterial bloom or a combination of both. Normally, filtration removes particles and the bacteria soon die off. Cloudy water in a mature aquarium indicates a serious pollution problem. Check that the filter is working properly, remove any rotting material (dead fish, uneaten food), and carry out a 30 percent water change every day until the water clears.

119 Protect your aquarium from dust and paint fumes

When redecorating, remember that many paints, varnishes, etc., give off harmful fumes, and tools such as drills and sanders produce dust that can get in the water and harm your fish. If moving the aquarium is impractical, seal it up with several layers of plastic wrap and move the air pump to a fume- and dust-free room. The fish will not suffocate and will be fine without food for a week or so.

120 Aerosol sprays can be lethal to aquarium fish

All aerosol sprays—not just insecticides, but polishes, window sprays and so forth—are potentially toxic to fish. The tiny droplets can enter the aquarium directly or via the air supply. Never use aerosols of any kind in a room with an aquarium. If you forget, a 50 percent water change and carbon filtration may—or may not—help. Prevention is better!

121 Make your own smoke filter for the aquarium

In a smoke-filled environment, you can avoid pumping tobacco fumes into the aquarium by making an air filter from a good-sized screwtop jar, half-filled with water. Drill two holes in the lid and silicone air lines in—the inlet pipe from the pump should extend below the water level in the jar, and the outlet to the tank should be almost flush with the lid. The air will bubble up through the water, and the fumes will go into the solution.

122 Aquatic plants are suited to certain environments

Wild aquatic plants have evolved in specific environments across the globe, and where one or more species is particularly favored by local conditions it quickly becomes dominant. One plant may benefit from the presence of another, such as shade-loving anubias, which might fall foul of slime algae if it were not shaded by a surface, light-loving plant, such as salvinia *(Salvinia auriculata)* or duckweed *(Lemna)*.

123 Obtain good-quality plants for long-term success

Much of your planting success will depend on where you source your stock, be it a mail-order supplier or a local store. If you obtain good-quality specimens to start with, they will travel better, establish more quickly and ultimately prosper in the aquarium, rather than fall into the category of wilted has-beens.

Tip 122 *In favorable conditions, aquatic plants soon establish.*

124 Keep newly arrived plants in good condition

Sometimes mail-order plants arrive sooner than expected. Rather than leave them in the wrapping to stagnate, always float them in another filtered aquarium until you are ready to plant them. Alternatively, set up a large bucket of mature, or at least dechlorinated, water and use an external or sponge filter to keep the water conditioned.

Tip 124 *Unpack mail-order plants promptly.*

125 Give plants a chance to settle into the aquarium

Bear in mind that no matter how much light and how many minerals are present, a change in conditions from the wild tropics to the aquarium is just too much for some plant species. The only way they can adapt to the new conditions is to shed leaves and regrow from the tips or rootstock. Water wisteria *(Hygrophila difformis)*, eusteralis *(Eusteralis stellata)* and some cryptocorynes often do this. Some take many weeks to regenerate, so unless a plant is obviously soft and rotting, be patient and give it a chance.

126 Wash plants before use to eliminate pesticides

Beware of suppliers that sell plants dry and straight from their import boxes. Pesticide use is common practice in Asia and elsewhere, and adding even low doses of residues to a tank via the plants can be lethal to aquarium fish. Always wash plants thoroughly in warm matured water before placing them in your display aquarium.

127 Check for good root growth when buying potted plants

When buying any potted plant variety, always look for signs of new roots developing through the sides and base of the basket. Beware of pots buried deep in the substrate, as the bases may be in poor condition. Any good retailer will be happy to sweep away some of the substrate so you can make a closer inspection.

Tip 127 *A healthy plant will show evidence of strong, new root growth through the container. Always remove pots before planting.*

128 Choose plants that are suited to your tap water

Rather than spend valuable funds on water treatments, try to select plant species that will be best suited to your local tap water conditions. Generally speaking, people in soft water areas do much better with cryptocorynes and calcium-sensitive species such as hairgrass *(Eleocharis acicularis)*. In chalky areas consider calcium lovers, including hornwort *(Ceratophyllum demersum)*, elodea *(Egeria densa)* and spatterdock (*Nuphar* spp.).

Tip 128 *Crypt* (Cryptocoryne walkeri *var.* lutea)

129 Buy a mature plant for a guaranteed display

When buying a plant from a retailer, it makes sense to invest in the more expensive mother plant, rather than a juvenile specimen, as it has probably been in stock longer and has had a chance to adapt to the local water conditions. Always check for healthy growth and signs of new shoots at the tips.

Better Aquarium Plants

130 Look for signs of life on bulb plants

When buying bulbs such as water lily (*Nymphaea* spp.), *Barclaya* species and aponogetons always look for signs of life. Little white stumps or the tips of leaves showing are good signs that suggest the plants are fresh imports. The bulb or rhizome is a sugar store ready to feed your new plant—when stimulated by the warmth of the aquarium, the plant will grow rapidly.

Tip 130 *The rhizome of orchid lily* (Barclaya longifolia).

131 Potluck plant selections allow you to experiment

When you set up a new aquarium you may need several plants to furnish it, without being quite sure which ones will thrive in your water conditions. Some shops and most mail-order companies operate an end-of-week or potluck collection of plants. Buying these enable you to pick up a whole range of species at low cost and experiment at leisure to see which of them are best suited to your conditions. You can then go back and invest the rest of your budget in the successful varieties.

132 Think about the proportions of your planted aquarium

The dimensions of the aquarium you choose are important for plant success. For example, it is no good setting up an aquarium more than 24 inches (60 cm) deep if you plan to use fluorescent bulbs as the light source. These provide effective illumination for a maximum water depth of about 18 inches (45 cm). Also think carefully about the width of the tank (the measurement from front to back). It needs to be generous to achieve a real sense of depth to the planting display. Ideally, select a tank measuring 36–48 inches (90–120 cm) long, 18–24 inches (45–60 cm) wide and 14–18 inches (35–45 cm) deep.

Tip 132 *A plant display suited to an 18-in. (45 cm) wide tank.*

Tip 135 *Adding a heater cable*

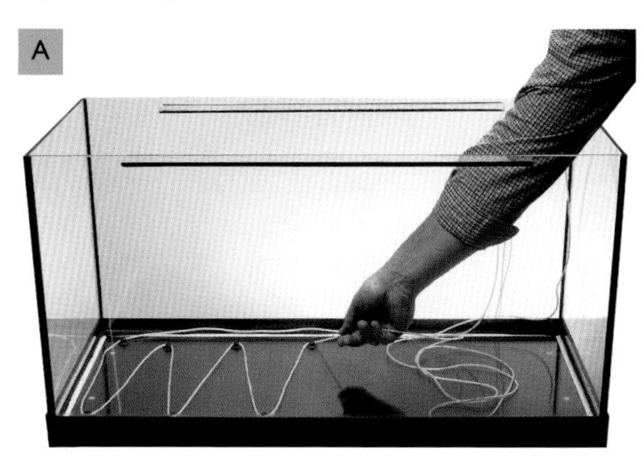

A *Install a heating cable evenly across the aquarium, guiding it around the rubber suckers provided.*

B *Using a jug, pour the washed substrate into the aquarium to a depth of about 3 in. (7.5 cm).*

133 Has your new plant been grown above water?

When you buy a plant such as a species of *Bacopa* you may not be sure whether it has been grown above or below the water (emerged or submerged). In this case, leave it to float for a week or so. If all but the top few leaves fall away, it means that it is the emerged form. Leave it alone, however, because the extra light and heat on the surface will stimulate aerial rooting and, with luck, the first underwater leaves.

134 Essential feeding for aquarium plants

A newly introduced plant will need to increase its chlorophyll content for greater photosynthetic efficiency. At this stage, it often requires a massive uptake of essential minerals, including magnesium (Mg), manganese (Mn), iron (Fe) and, of course, nitrogen (N). Be sure to use a long-term or substrate fertilizer from the start and one rich in iron, magnesium and manganese. It should not contain added nitrates or phosphates. Do not confuse phosphate (PO_4) with useful potassium (K).

135 The benefits of heating the substrate

If you cannot, or do not wish to, use undergravel filtration, a heater cable (50-watt or greater) will have a beneficial effect in the aquarium. It will help to prevent long-term chilling and stagnation of the substrate, which leads to general degradation and foul gases. However, while underfloor heaters help, they can be difficult to find and seldom produce enough convection to prevent stagnation. A little extra spring cleaning with a gravel cleaner is always necessary. (See also Tips 53 and 434.)

136 Nutrient-rich substrate will give you a flourishing display

It is generally more cost effective to invest in a good-quality substrate, as the clay-based types (usually made of laterite) are rich in iron and, being porous, can retain many more beneficial filter bacteria than standard aquarium gravel. Alternatively, you can add a seam of nutrient-rich laterite between gravel layers. Money spent at the outset on providing the best growing medium will more than pay dividends long term.

Better Aquarium Plants

137 Feeding aquarium plants the natural way

If you own a guinea pig, hamster, gerbil or rabbit you can use the droppings to feed your aquarium plants. Aquarists were using this fertilizer successfully long before expensive, manufactured fertilizer pellets were available. Simply push a dropping into the substrate close to the roots of each plant. Repeat at intervals of three to six months. Do not overdo it; three to four droppings per 26 gallons (100 L) of water will suffice. Add a dose of a mild bactericide to prevent any infections arising.

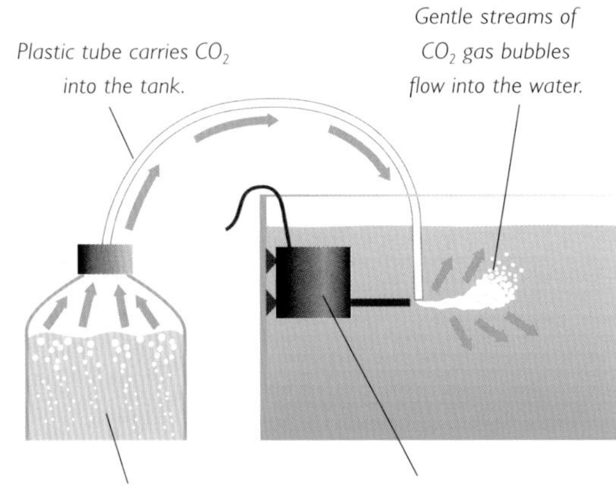

Tip 138 *A simple CO_2 system.*

138 Carbon dioxide promotes healthy plant growth

Adding carbon dioxide (CO_2) to the aquarium is beneficial, as it is the main way that plants obtain their carbon during photosynthesis. It will speed growth in many species, especially cryptocorynes and floating plants. You do not need to spend a great deal of money on a CO_2 system, as there are several reasonably priced hobby kits on the market. The gas bottle replacements for more elaborate systems are also inexpensive.

139 Use a water conditioner to avoid chlorine strip

Do not think that by simply waiting a few days for the chlorine in tap water to dissipate it will be safe to put plants into the aquarium. Always use a good-quality water conditioner. Vals and cryptocorynes are very prone to chlorine strip, whereby the cell membranes are ruptured, causing a catastrophic structural collapse. This is usually the main cause of cryptocoryne rot, rather than an infective agent.

140 Do not bury plant bulbs in the substrate

Never bury bulb species in the substrate when planting them. Bulbs are stimulated by heat and light, and the anaerobic microclimate in the substrate will encourage molds—the entire bulb will rot before you see a single leaf. Just make a dent in the substrate for the bulb and within days new root growth will hold it in place.

141 Lighting levels for aquarium plants

As there is a considerable drop in available light in the captive environment compared to the outdoors, be sure to use at least two fluorescent bulbs with different spectrums, especially if the width of the aquarium exceeds 12 inches (30 cm). A rough guide is 10–15 watts per square foot (900 cm^2), but aquarium plants vary in their lighting needs so check the advice given with each plant. Remember that some "marine" tubes provide UV light and are ideal for plant use too. (See also Tip 66.)

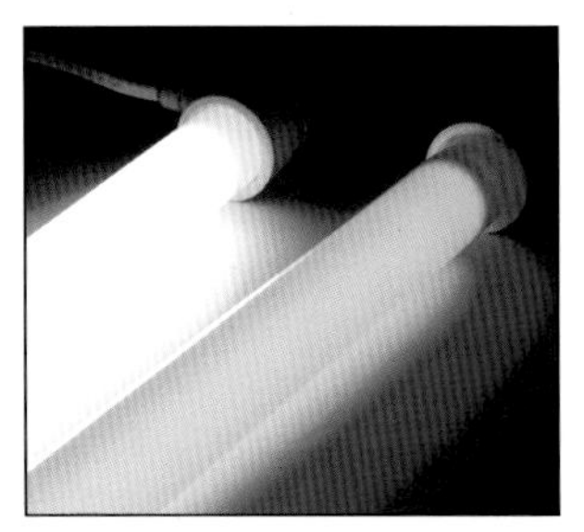

Tip 141 *White and actinic light bulbs.*

142 Use a hairnet to fix plants until they root

To secure plants such as Java moss *(Vesicularia dubayana)* or pellia *(Monosolenium tenerum)* to an odd-shaped piece of wood or tufa rock, try using a dark-colored hairnet to hold the plant securely without suffocating it. After a few weeks, the plant will root onto the decor and the net can be removed.

Tip 142 *Pellia* (M. tenerum) *netted to a plant stone.*

143 Use lighting to create the illusion of depth

When you illuminate your aquarium plants, always keep the brightest bulb at the front so the reflected light shines back off your plants. This will enable you to appreciate their colors at their best. Adding a blue bulb toward the back not only improves UV for the plants, but adds a feeling of depth by casting shadows in the leaves.

144 Bring aquarium plants up toward the light

Some red plants need an increased amount of light, but may be too short to reach the surface regions. Use a plastic twist tie to secure them to the upper parts of a robust twiggy plant, such as *Hygrophila corymbosa* 'Stricta', which can give them a piggyback. The attached plants will develop aerial roots but will absorb most of their nutrients through their leaves. The result may look strange to start with, but if you choose two similar plants, they will appear to become one after a time, almost like a tree graft. Trim the aerial roots if they look unsightly.

145 Use plants as a signal of excessive light levels

Certain species, such as hornwort *(Ceratophyllum demersum)* and *Hygrophila* 'Rosanervig', turn red or pink in high light conditions. It means they are producing flavonoids, a natural response to shield the plant tissues from excess sunlight. This is a signal that your aquarium lighting is strong and likely to encourage unwanted algae bloom or damage to shade-loving plants such as anubias.

Tip 145 Hygrophila *'Rosanervig' develops red leaves in bright light.*

Better Aquarium Plants

146 How to encourage dry leaves in the aquarium

Shallow-water tanks for amphibians and mudskippers, in which an area is set aside as a marshlike habitat, are becoming increasingly popular. As a change from sphagnum moss, try acclimatizing some aquatic plants, such as ludwigias, Amazon swords (*Echinodorus* spp.) and hygrophilas, so they can form dry leaves. This is easy to do by laying the submersed plant next to the muddy area so that new aerial roots can take hold. Dry leaves will soon emerge at the top. Just like land plants, they will obtain atmospheric CO_2 and grow very rapidly.

147 How do I keep the water clear in a plant-only setup?

If you do not intend to keep fish but simply want to see plants growing, you can achieve excellent results using an unfiltered tank that receives some natural daylight and has artificial light to extend daylength in winter. Under such conditions, free-swimming bacteria and algae will soon be a problem, but the trick is to buy a bag of *Daphnia* (water fleas) and put them in the tank three or four days later (when the tap water in the bag has lost its chlorine). They will consume millions of algae and bacteria, keeping the water crystal clear. Given the increased levels of CO_2 present in these conditions, the plants should grow rapidly.

Tip 147 Daphnia *keep the water clear in a plant-only tank.*

148 Let plant leaves form the main display

With a few exceptions, aquarium plants that grow from a corm or bulb, such as aponogetons and *Barclaya* species, are best cultivated for their attractive leaves alone. It is important not to let them flower, as they will set seeds that will not germinate in the tank, thus wasting all the energy stored in the bulb. In many cases, the bulb and plant will die off within weeks of flowering. Simply pinch out any flower spikes before they break the water surface.

Tip 149 Vesicularia *'Christmas' is an attractive aquarium moss.*

149 Give mosses a chance to show their best

Mosses often fail because they are not treated correctly. Left on the substrate they act like mops, picking up debris that will ultimately suffocate them by starving them of the light they need for photosynthesis. Instead, tie them to rocks or bogwood at a higher level. In common with riccias, they actively absorb passing minerals, so place them in a gentle stream of water from the powerhead or filter. This will also keep them clean.

Tip 150 *Trim off wayward stems to keep the display in good shape.*

150 Choose the right strategy when pruning aquarium plants

From time to time, plants will need pruning. First establish where the new growth is arising. If it is from a central crown, as in vals and Amazon swords, remove entire outside leaves to reduce the height rather than risk damaging the crown. In the case of twiggy or singled-stem species, such as hygrophila, elodea *(Egeria densa)* and cabomba, simply crop at the base of the plant to just below where any aerial roots are showing to achieve the required height reduction. Replant trimmings to fill gaps elsewhere.

151 Drastic pruning can spur plant growth

Plants such as Amazon swords (*Echinodorus* spp.) can be pruned back hard by removing nearly half of the outer leaves. The shock will stimulate the plant to put down strong roots as an insurance policy, allowing it to recover when the perceived emergency has passed. You can keep such plants compact by regularly removing the older leaves before they reach full size.

152 Regular thinning keeps plants in good condition

Remember that the aquatic garden you have created needs to be kept in balance. Cut back invasive plants and remove dead leaves, as they are a time bomb ready to release phosphate back into the system. Do not be tempted to kill plants with kindness by doubling the dose of iron—without sufficient light or CO_2, this will simply give algae the chance they have been waiting for to pounce.

153 Propagating plants at the top of the tank

Most aquariums have strengthening bars and surrounds that can support small clay pots filled with aquatic growing "soil." It is a fairly simple process to propagate new Amazon sword species (*Echinodorus* spp.) from the baby plants that arise from the flower spikes. In favorable conditions during the summer, these periodically reach out above the surface. Peg one daughter plant into each pot, keep the soil wet at all times and the roots will begin to appear in days.

154 Aquatic plant waste makes good houseplant compost

When trimming old, decaying leaves or those covered in algae, consider using them as houseplant compost. Similarly, when you rinse cleaning scourers after scraping algae from cover glasses or the sides of the aquarium, keep any green water. Both are rich in nitrates and phosphates. Harvesting algae in this way not only improves the aquarium water quality, but also benefits your non-aquatic plants.

155 A filter system that suits both fish and plants

You will need filtration in a planted tank, especially if it includes fish. In this situation, a good compromise is to use an external filter coupled with a reverse flow undergravel filter. In this system, instead of water being drawn up the uplift tube by a powerhead, the clean, warm outflow water from the external canister filter flows down the uplift tube to the space beneath the filter plate. It then flows upward through the gravel bed. Although it is true that undergravel filters slow plant growth a little, they do provide the most cost-effective option to maintain a stable aquarium environment. To ensure sufficient oxygenation for the fish at night, when no oxygen will be coming off the plants, operate an air curtain or airstone.

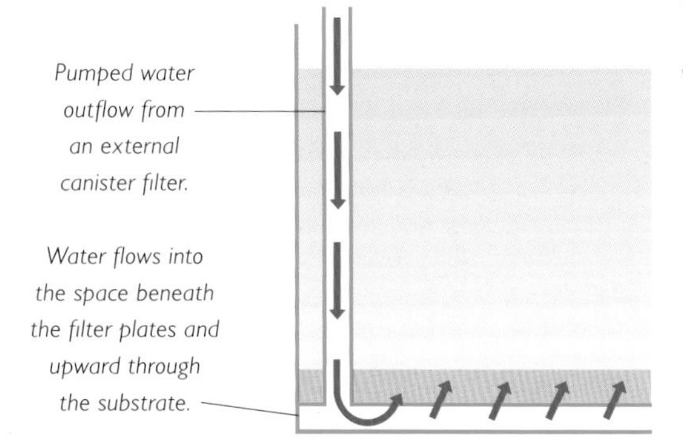

Tip 155 *Reverse-flow undergravel filtration.*

156 Setting up a miniature reed bed filter above the tank

If you have an external filter with a spraybar, you can set up a miniature reed bed filter, fed by the continuous flow of nutrient-rich water as it returns to the tank. This will feed marsh species, such as acorus rushes *(Acorus gramineus)* and certain woody-stemmed plants, such as giant hygro *(Hygrophila corymbosa)*. Use the glazing bars around the top of the aquarium to support this filter. It can be made out of a small plastic seed tray, or similar, but do not exceed about 2 pounds (1 kg) in weight to avoid damaging the tank. The result will look like a terrestrial extension to the aquarium plant display and also reduce levels of phosphates and nitrates in the tank.

Tip 157 Hornwort (C. demersum) *helps to control phosphates.*

157 Keep phosphate levels down by using plants

Phosphate in excess of 4 mg/gallon (1 mg/l) quickly encourages algae to flourish in the aquarium. Rather than resort to chemical absorbents, try using fast-growing plants, such as hornwort *(Ceratophyllum demersum)*, dwarf hygro *(Hygrophila polysperma)* and sea grass *(Heteranthera zosterifolia)* to keep phosphates at around 1.2–2.5 mg/gallon (0.3–0.5 mg/l). Healthy plant growth is both nitrate- and phosphate-dependent, so pure reverse osmosis water is of little use in a healthy planted aquarium. (See also Tip 107.)

158 What is the brown slime on some of my plant leaves?

If you feed your fish fresh or frozen protein-rich foods, the water may become slightly cloudy, thus reducing vital light transmission to your plants. Diatoms also thrive in protein waste and can cover the leaves in a brown slime. Keep some algae loaches *(Gyrinocheilus aymonieri)* or otocinclus to gently rasp off the brown film and, better still, reduce the amount of food you offer your fish.

Tip 158 *Otocinclus helps to keep plant leaves clean.*

159 Propagate extra supplies of plants in a glass jar

Sometimes it is too difficult or expensive to obtain sufficient numbers of plants such as riccia or mini hairgrass *(Eleocharis parvula)* to create an instant effect in the aquarium. However, you can increase your stocks using a large glass jar set up on a windowsill that does not receive direct sunlight. Place the plants in matured tank water and add about 1½ ounces (50 ml) of carbonated springwater every day to supply carbon dioxide. Feed the plants with a couple of drops of houseplant food per week. Do not allow the jar to roast in the sun or you'll end up with pea soup!

160 Protecting your plants from herbivorous fish

Some plants, such as anubias, are tough enough to withstand herbivorous fish. If you have plant-grazing fish, such as plecostomus (*Peckoltia* spp.), you can protect plants by using diversionary tactics. Simply offer the fish something better to eat, such as scalded young leaf lettuce, a slice of zucchini or one of the many good-quality vegetarian tablet foods available, such as algae wafers. As many plant-grazers are nocturnal, it is best to add these treats just before lights out. (See also Tips 277 and 294.)

Tip 160 *Clown plec* (Peckoltia vittata) *enjoys a diet rich in green foods.*

Better Aquarium Plants

161 Plants can be affected by fish disease treatments

If you suffer an outbreak of fish disease, such as ich or velvet, in the aquarium, remember the plants before you add copper-based remedies. Move the delicate plant species to a quarantine tank and treat them with a gentle, broad-spectrum bactericide because they could carry infective spores back into the main tank after the fish treatment has finished.

162 Protect local waterways from escaped plants

Bear in mind that most of the plants bought for aquarium use are tropical in origin and will not survive winter conditions outdoors in temperate climates. However, a few species, such as some of the pennyworts (*Hydrocotyle* spp.) milfoils (*Myriophyllum* spp.) and stonecrops (*Sedum* spp.), can survive outdoors, especially in milder conditions. Always burn or compost aquarium waste; never flush it into the sewers or put it into local waterways. The plants may outcompete native species and cost a great deal of money to control. If we cannot be responsible in this way, then many valuable aquarium species risk being lost as the result of an import ban.

Tip 163 *Snails breed rapidly and can damage plant leaves.*

163 Tackling snails in the display aquarium

Snails can be troublesome in your aquatic garden. The small, brown ear-shelled snails are the worst offenders, as they quickly overpopulate, punch little holes in plant leaves and, in a severe infestation, obscure your view through the glass. Do not tackle the problem with irritants or copper-based remedies, as these strip the protective outer skin from many plants, leaving them open to bacterial attack and so-called meltdown syndrome. Simply buy a couple of clown loaches (*Botia macracantha*) measuring 2 inches (5 cm) or more. Any *Botia* species will do, but clown loaches are often the most peaceful and attractive option. After worms, snails are their favorite food. However, some snail species such as the Malayan mud snail or coneshell, which eat detritus in the substrate and the ramshorns that graze slime algae, can be beneficial if the population is kept in check.

Tip 163 *Damage from the center outward indicates snail infestation.*

164 Aphids can attack your aquarium plants too

Another occasional pest—mainly seen on floating plants or leaves—is the reddish brown house aphid, the type that fly in the window in spring and set up home on your indoor plants. They suck the sap of juicy plants and can reproduce at frightening speed. For obvious reasons, you cannot use any pesticides in the aquarium, so the best method is to remove the affected plant carefully and wash it in a tub of very weak soapy water for a few seconds. Then rinse it and put it back in the tank, taking care to use a fine net to capture any floating bugs you may have missed.

Tip 165 *Slime algae on the fronds of an aquatic fern.*

165 Coping with slime algae and molds

Slime algae and fungal molds appear in many forms. Blue-green algae can be toxic and may occur during periods of high summer temperatures, when the tank reaches 90°F (32°C) or more. Diatoms are free-swimming algal cells, often present in underlit or new aquariums that have not matured. In many cases, the presence of some algae-eating fish to rasp them away is enough to cure the problem. Molds thrive in dirty, neglected tanks. Initially they feed on dead matter, but they can spread to healthy plant tissue. Some carbon filtration, or chemical absorbent in severe cases, along with a good cleanup is usually all that is required to rectify the situation.

166 Avoid introducing green filament algae into the tank

The more troublesome filament algae are often introduced on bogwood dug out of the soil or, very possibly, on some types of frozen food where the spores have survived the irradiation process used to kill bacteria. Bloodworm, daphnia and tubifex should all be washed through a net before being fed to fish. However, it is often a matter of trial and error.

167 Do not move large plants when cleaning the aquarium

Try to avoid moving a specimen plant when cleaning, as it will have a global root network, twice the size of any visible growth. An elderly plant may not tolerate the disruption.

168 How to move a large plant if you have to

If you have to move a much-loved specimen to a different aquarium, treat it as you would a mature garden plant. Remove all the loose substrate until you can feel the major roots. Then slide a large, fine net underneath and lift out the plant, keeping the original substrate around the rootball, and transplant it into its new home. Do not wash away all the debris from around the roots; the filter system will deal with that later.

169 Sit back, relax and enjoy the show

You can make two big mistakes with aquarium plants: one is gross neglect, resulting in a compost heap or out-of-control jungle, and the other is to fiddle when things are going just fine anyway. There is no point in having a relaxing hobby if it is going to become a source of constant worry. Take time to just sit and enjoy the aquarium, and change your viewpoint slightly every day.

Better Aquarium Plants

170 Helping plants that feed through their leaves

Some plants, such as Java fern *(Microsorium)*, take in the majority of their nutrients through their leaves and will require a constant, gentle flow of water to move waterborne nutrients around them. Positioning these plants near a filter outlet or deliberately positioning small pumps around the aquarium will help to achieve this.

Tip 170 *A finely branched Java fern (Microsorium spp.).*

171 A limited range of plants can make a big impact

A planted aquarium does not have to include a wide range of species; sometimes a great display can be created with just a few different plants. Good candidates for a planted species tank include val *(Vescularia dubayana)*, cryptocorynes, Amazon swords (*Echinodorus* spp.) and hygros (*Hygrophila* spp.). These plants all look great in large groups and can be left to overrun the aquarium.

172 Making the most of carbon dioxide fertilization

If you use a carbon dioxide unit that can be controlled electrically, set it on a timer so it will come on for only the middle hours of the day. During the first few hours of light there is enough carbon dioxide in the water from fish and plant respiration, while for the last few hours of light the plants will use up the remaining carbon dioxide. If your lights are on for 12 hours a day, you only need 6–8 hours of carbon dioxide, and reducing carbon dioxide output will save you money and keep your aquarium more stable.

173 Hardy plant species for low-light aquarium conditions

Many aquarium plants are suited to lower light conditions and will thrive in a low-tech aquarium. Hardy species such as cryptocorynes, anubias and Java ferns (*Microsorium* spp.) will thrive with minimal maintenance, simple fluorescent lighting and no extra carbon dioxide. Just provide a good-quality, nutrient-rich substrate and they will get everything else they need from the waste produced by your fish.

Tip 173 *Cryptocorynes thrive in low-light conditions.*

174 Housecleaning shrimps will keep your plants clean

Japonica shrimps *(Caridina japonica)* are an excellent addition to a planted aquarium. A group of these shrimps will constantly remove and process algae and debris, keeping your plants in top condition and free from detrimental algae. For fine-leaved plants such as cabomba and mosses, these shrimps are just about the only thing that will prevent debris from settling and blocking light to the leaves.

175 Floating plants add another layer to the tank display

Floating plants add an extra element to a display and are excellent for removing excess nutrients from the water. To keep floating plants successfully you should make sure that the aquarium receives adequate ventilation above the surface. If you have a closed hood with just a few openings, a small fan can be positioned to create a gentle airflow through the hood. Good floating plants for the aquarium include salvinia *(Salvinia auriculata)*, water lettuce *(Pistia)*, riccia, duckweed *(Lemna)* and *Limnobium* species.

Tip 175 Limnobium laevigatum, *a waxy-leaved floating plant.*

176 Make a feature of specimen plants

Specimen plants are an excellent way to add a centerpiece feature to a display. Dominant plants such as tropical lilies, object-rooting plants (like anubias, Java ferns and *Bolbitis* spp.), larger Amazon swords and large aponogetons all look good when placed between smaller plants. Make sure you allow adequate room for these plants to grow, or they could soon crowd the aquarium.

Tip 176 *White Egyptian lotus* (Nymphaea lotus) *leaves provide a stunning aquarium feature.*

Better Aquascaping

177 I'd like to aquascape my tank. Where do I start?

Always plan your aquascape before you construct it, even if you think you know what you want to create. Drawing diagrams, sketching the display and making shopping lists often help to highlight any potential problems and provide solutions that you may not have thought of initially. Why not show your ideas and sketches to fellow aquarists or your retailer? They may come up with useful thoughts and suggestions.

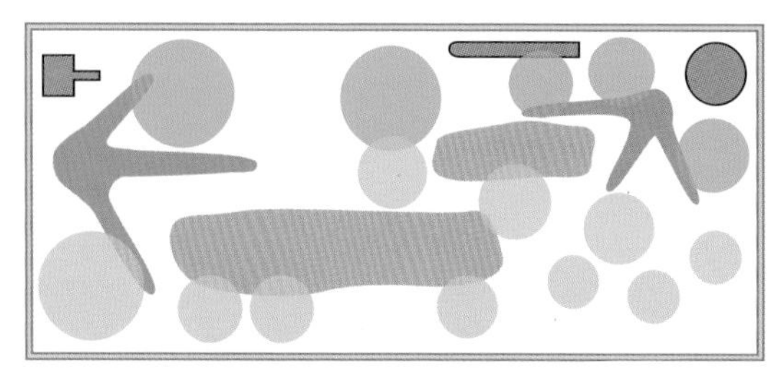

Tip 177 *Draw up a simple plan of your aquascape.*

178 What looks best—a formal or an informal design?

Aquascapes can be formal or informal in design, but to achieve a natural appearance aim for a mixture of both. Plan and set up your aquarium in an organized way and then add some informal final touches. Place single plants slightly away from groups of the same species and scatter mixtures of different gravels, small stones and pieces of wood across the aquarium floor.

179 It can help to take inspiration from nature

Aquascaping is a matter of personal taste, but for the best results take your inspiration from nature. Try recreating a natural biotope (i.e. an area of nature) such as a swamp, riverbank, fast-flowing stream, jungle brook, lake or estuary. Complement the design of the aquarium by choosing a selection of suitable fish from the same natural environment.

180 Your fish should feel at home in their aquarium

Always try to match your aquascape to your fish or vice-versa. Heavily planted aquariums make good homes for swamp-dwelling fish, while aquariums with minimal decor would be suited to large, open-water swimmers. Fast-flowing aquariums with gravelly bottoms would be a good environment for stream-dwelling fish.

Tip 180 *Pentazona barbs* (Barbus pentazona) *are interested in their surroundings.*

Tip 179 *An aquascape modeled on a West-African streambed.*

181 Fish are part of the aquascape too

Fish are the moving element in an aquascape and so should be considered as a major part of the display. Think carefully about the species you choose. Do you want bright, boldly shaped fish such as gouramis and rainbowfish; fast, darting fish such as barbs and danios; or slow-moving fish such as discus and angelfish? For added interest, add a few fish that you are unlikely to see all the time, such as some small algae-eaters or nocturnal species.

Glowlight tetra (Hemigrammus erythrozonus)

Neon tetra (Paracheirodon innesi)

Tip 181 *Tetras add a flash of color.*

182 I can't find the right plants for my biotope aquarium

A "true" biotope aquarium contains only plants and fish from the same area in nature. In the case of plants this may be difficult, as many naturally occurring plant species are not available or do not grow well in aquariums. However, the range of available aquatic plants does cover almost all shapes, styles and colors, so using similar plants from outside the biotope range will still create a display that's true to nature.

183 Choose the appropriate aquarium background

Use an aquarium background to hide any unsightly cables, pipework and wallpaper that are visible through the rear glass of the tank. Choose your background with care; for the overall look of the display, a simple, all-black background often works best. Alternatively, a number of very realistic three-dimensional backgrounds are available, although some can be quite expensive.

184 Mix and match substrates for a natural appearance

For a more natural substrate appearance, use several different substrates mixed together. For types of a gravelly streambed, use different grades of pea gravel, along with small, medium and large pebbles and cobbles. Other aquariums may benefit from a sprinkling of dark or subtly colored gravels.

Lime-free gravel

Mixed-grade pea gravel

Red chippings

Black quartz

185 How do I stop my fish from disturbing the substrate?

When layering different grades (sizes) of substrate or when keeping fish that like to dig, use a gravel tidy. It can help to keep the substrate in place and allows you to reposition the top layer easily. A gravel tidy is simply a flexible mesh that can be positioned within the substrate, preventing the layers above and below the mesh from mixing. Plant roots can easily grow through the mesh.

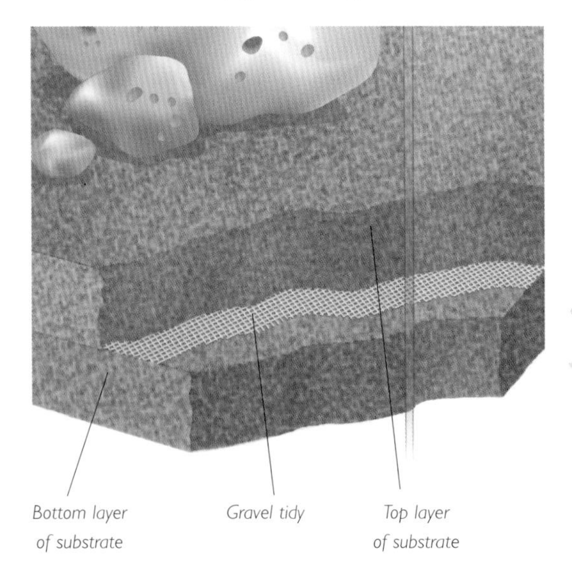

Tip 185 *Gravel tidies prevent root disturbance in the lower substrate.*

186 How can I stop a sand substrate from turning black?

Sand is a popular choice for many displays, but it can compact and stagnate over time, turning black and releasing toxins. Plant roots will help to stop this from happening, but in open areas you will need to agitate the sand gently with your fingers at least once or twice a week. Heating cables placed beneath a sandy substrate can also help to keep the area oxygenated. (See also Tip 135.)

Tip 187 *Catfish will grub around in the substrate.*

187 Choose safe substrates for bottom-dwelling fish

If you are keeping scavenging fish, such as bottom-dwelling loaches and catfish, avoid sharp or jagged rocks and gravels. These fish can easily damage themselves on unsuitable decor and succumb to bacterial infections. Some of the more delicate fish can only be kept properly with a fine, sandy substrate through which they can sift and into which they may even bury themselves on occasion.

188 Will a dark substrate make my display look gloomy?

Quite the opposite—darker substrates can have a striking effect, actually showing off plants, fish and decor better than lighter substrates. It is often worth looking through the colored gravels to find the more subtle browns, grays, dark reds and blacks. These can be used to create a darker and slightly different effect. Furthermore, many fish find a darker substrate more relaxing, resulting in lower stress levels and better overall health and color.

Tip 186 *Gently disturb sandy substrates to avoid compaction.*

189 Don't be afraid to try unusual (but safe) decor

For a cheap and safe source of unusual rock decor, try some normal coal. Once the pieces have been washed and are free from dust, they are perfectly safe to use in the aquarium. Dark coal pieces can have quite a dramatic effect and contrast well against plants and fish.

Tip 189 *Pieces of coal look striking.*

190 Create families of decor with bits and pieces

For a more natural display, break up a few pieces of rock and scatter the small broken parts around the base of larger rocks of the same type. Adopt the same strategy with pieces of wood, complementing a larger piece with smaller fragments.

Large angular pieces of slate

Tip 190 *Create a natural display by using slate chippings around the base.*

191 Protect the aquarium from damage by sharp rocks

When creating a large rockscape, it is a good idea to place the rocks on a supporting structure to prevent contact with the aquarium base and deter fish from digging underneath. Use a sheet of criss-crossed plastic (often called egg crate), an undergravel filter plate or a sheet of Styrofoam. Any of these will protect the bottom glass from damage caused by sharp points or ridges on any large rocks.

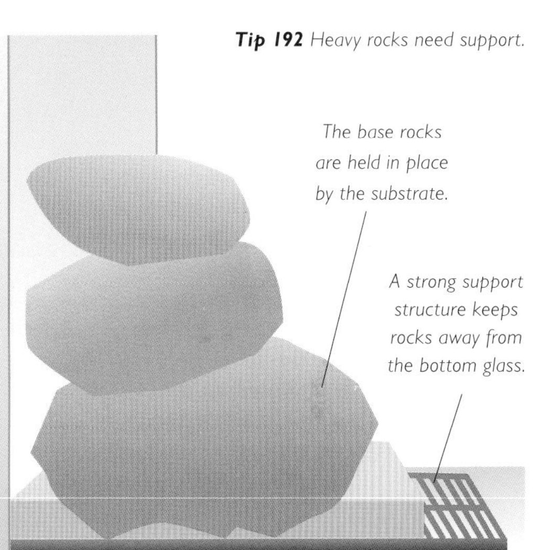

Tip 192 *Heavy rocks need support.*

192 Building up secure rockscapes and backgrounds

Backgrounds made up of several rocks or other decor should be fixed in place with silicone before you fill the aquarium. To prevent the background or rockscape collapsing during construction, place the empty aquarium on its back (ideally on some foam or other suitable support), with the top slightly raised. The decor can now be positioned and secured horizontally. When the silicone has dried, turn the aquarium the right way up ready for filling.

Better Aquascaping

193 Lava rock is safe to use without fixing

A rockscape can be difficult to adjust once it is fixed in place, but if it is not secured it can become a safety hazard. However, you can build up a non-fixed rockscape using lava rock, which is very light and easy to shape so that pieces lock together. If this type of rockscape should ever collapse, the lava rock is so light that it should not damage the tank, with the exception of a few scratches.

Tip 193 *Find pieces of lava rock that fit together.*

194 Use aquarium sealant to create decorative features

Many adhesives can only be used properly when they are applied to a dry surface. To repair, add or construct objects underwater, use a waterproof, silicone-based adhesive that will set and dry underwater. Don't worry about a messy appearance; once dried, the excess can be removed and time will weather the sealant until it is hardly noticeable.

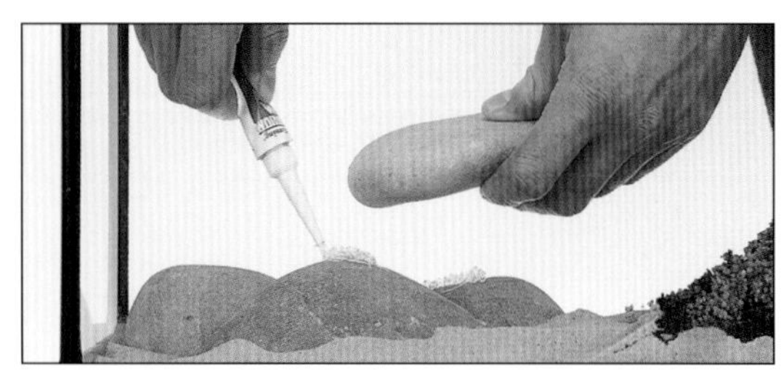

Tip 194 *Squeeze large beads of silicone onto a clean surface.*

195 Carry out the fizz test to make sure rocks are safe

Some gravel and rocks are calciferous, meaning they may release calcium compounds that increase the water's hardness. To test these rocks, simply place some vinegar on the surface and look for any gentle fizzing. If the rock or gravel fizzes it is only suitable for fish that prefer very hard water, such as Rift Lake cichlids or certain rainbowfish. Some rocks need a stronger acid to give a reaction. Try rust-removing liquid or gel, which contains phosphoric acid. Always wear gloves when using stronger acids and thoroughly clean any tested rockwork before using it in the aquarium.

Tip 195 *Make sure the rock is suitable for the aquarium.*

196 Can I collect rocks from the wild to decorate my tank?

No, never collect rocks from the wild; always use rocks obtained or recommended by an aquatic retailer. Many rocks that may seem safe often contain heavy metals and toxins that will leach into the aquarium water. The small volume of water in an aquarium allows these pollutants to reach dangerous levels very quickly, harming plants and fish.

197 Different fish prefer different rock types

Angelfish *(P. scalare)*, discus *(S. aequifasciatus)* and dwarf cichlids like flat rocks, such as slate, on which they can lay eggs. Rift Lake cichlids prefer stacked rocks that create lots of gaps, such as rounded cobbles or odd-shaped ocean or lava rock. Loaches and catfish will do best with smooth rocks that they can snugly hide beneath without damaging themselves.

198 The water in my tank looks brown, what should I do?

Bogwood can release tannins into the water that will turn it a light yellow tea color. Although this does acidify the water slightly, it is not harmful to the fish, and it may even provide more natural and preferred conditions for some fish. For an Amazonian pool design, the darker water will add a realistic element to the display. If you do want your water to return to its original clarity, use a chemical filtration medium, such as activated carbon, to remove the tannins.

Tip 198 *An Amazon pool aquascape with tinted water.*

199 Do I need to soak bogwood bought from a store?

When you buy bogwood, always ask if it has been presoaked; in most cases it will not have been and you will need to treat it to release the tannins. Soaking will also help to make it less buoyant in the aquarium. The process will take up to two weeks, so allow time before you begin setting up. Tannins may continue to leach into the water, causing a drop in the pH level, so carry out regular water changes.

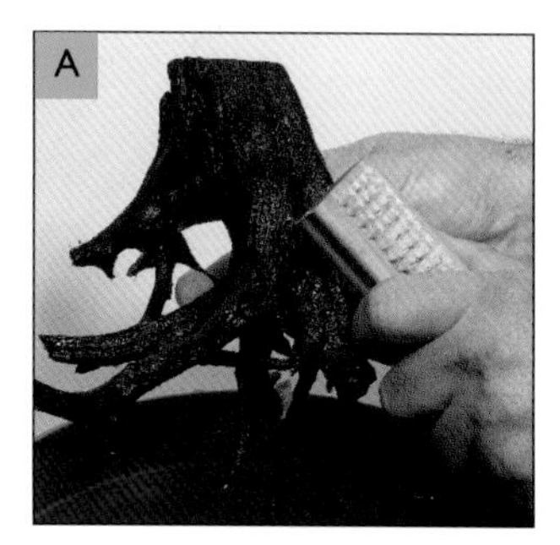

A Start by brushing the bogwood to remove dirt and debris. Wet it to remove stubborn marks and use a small brush to get into the narrow crevices.

B Keep the wood submerged. When the water looks like weak coffee, replace it with fresh tap water. Repeat the process as often as necessary to remove most of the tannins.

C It may take up to two weeks and several changes of tap water before the water remains clear. At this point, you can place the clean, waterlogged bogwood into the aquarium.

200 Is it safe to put bamboo canes directly into my tank?

Bamboo can be used to create great visual effects in the right display, but if not prepared correctly it can rot very quickly and cause harmful bacterial and fungal blooms. Soak thin canes for a week before use, and remove the skin (a thin membrane layer inside the wood) from larger pieces. Soak these for several weeks, or coat them with a clear polyurethane varnish.

Tip 200 *Coat large pieces of bamboo with a suitable varnish.*

201 What can I do to make bamboo less buoyant?

Bamboo is incredibly buoyant, even after soaking, making it tricky to position in the aquarium. You can cut pieces so that they fit snugly between the substrate and the glass bracing bars around the top edge of the aquarium. You can also silicone pieces to rocks, slate or pieces of glass placed beneath the substrate. Fill large bamboo pieces with some heavy substrate or cobbles to weigh them down.

Tip 202 *Brushwood must be completely dead and dry before use.*

202 Is it safe to use dead twigs in an aquarium?

You can use thin, twiglike brushwood—a great free source of realistic decor—but it is important to make sure there are no living parts left, which will rot and pollute the aquarium. To check this, simply break open the thickest part; if it bends rather than snaps or contains any green material, it is still living and should not be used. Soaking the wood before use will remove any nasties.

203 Overhanging branches add a natural dimension

To add an extra dimension to a display, fix pieces of decor, such as twisted roots or branches, to the aquarium hood with cable ties. They will then dangle into the aquarium from the surface without reaching the substrate, appearing like overhanging branches. For an even more striking effect, try growing mosses on these overhanging branches.

204 Are there advantages to using artificial decor?

Artificial rocks and wood are usually hollowed and make great hiding spots for fish, often better than the real thing. Even if you use natural materials, a few well-matched artificial pieces will soon blend in once they have become a little worn. They will provide great hidey holes for your fish.

Tip 204 *Artificial decor has many benefits.*

This fake bark is a good shape for hiding internal filters. The holes also provide shelter for fish.

Artificial wood looks natural and does not alter the color of the water.

Synthetic rock is inert and safe for aquariums.

Fake wood is available in manageable sizes.

205 How can I clean pieces of artificial decor?

You can clean artificial decor, rocks and synthetic plants with a specially made aquarium cleaning treatment. Take the pieces out of the aquarium and soak them overnight in water containing the cleaning solution. In the morning, rinse and return the pieces to the aquarium. Only use chemicals in the aquarium, or on aquarium equipment, that are specifically designed for aquarium use.

206 How can I conceal the equipment in my aquarium?

You can use carefully placed decor to hide aquarium equipment such as filters and pipework. Make sure that any equipment hidden by decor is easily accessible for maintenance and that heaterstats are positioned in an area of water movement for an even distribution of heat. Only use lighter decor around heaterstats to avoid breakages, or use a heater-guard.

Tip 206 *Choose decor shaped to hide tank equipment.*

207 Protect your heaterstats from rockwork

Heaterstats are breakable and require unobstructed surroundings to distribute heat, so they are difficult to hide in heavily rock-based designs. A good solution is to use heater-guards, which are readily available and fit most heaters. The guard will stop any rocks from touching the heater and create a water-filled gap between the rocks and the heater.

Tip 207 *Heater-guards also shield fish from burns if they shelter too close.*

208 Mass planting achieves a greater impact

Plants often grow in dense patches in nature and using a large number of only a few species, rather than smaller numbers of many different kinds of plants, can create an impressive aquarium. For the best effect, try using shorter plants, such as small cryptocoryne species, in the foreground, and build up the background with tall stem plants with small leaves or tall plants with leaves growing directly from the base.

Tip 208 *Closely planted cryptocorynes create an attractive feature.*

209 High-rise plants that grow on wood or porous rock

Some plants, such as mosses, anubias, Java ferns (*Microsorium* spp.) and *Bolbitis* species, will happily grow on solid objects rather than in the substrate. To achieve this, remove any growing medium (such as rockwool), trim the roots and attach the plant to a porous rock or piece of bogwood using black thread. The thread will eventually break down and the plant's new roots will keep it firmly attached.

210 Create a plant tree on some twisted roots

Try using long, thin, twisted roots planted with anubias, and *Bolbitis* species and Java ferns (*Microsorium* spp.) to create a plant tree. The exposed roots can be covered with mosses such as Java moss *(Vescularia dubayana)*. This method of planting is particularly good for streamlike displays where the pea gravel and pebble substrate is not suited to most plants.

Attaching a plant to wood

211 Choose the best plants for spawning fish

Some fish like to spawn on specific plant types, so try to include these in your aquascape. Angelfish and some catfish will spawn on large leaves, such as those of the Amazon swords (*Echinodorus* spp.). Egg-scatterers, such as rainbowfish, like to spawn above Java moss *(Vescularia dubayana)* or dense thickets of similar vegetation. Asian barbs will spawn among the finely divided foliage of cabomba or *Limnophila* spp.

Tip 211 *Cherry barbs* (Barbus titteya) *spawning.*

212 Where do I find plants for a paludarium?

It can be difficult to obtain suitable plants for paludariums (aquariums with plants rooted above the waterline). A good source is the large number of marginal plants sold for ponds, and as these are sold in larger pots than the aquarium plant equivalents, you get far more for your money. A bit of trial and error may be required to find the right species. Plants such as *Lysimachia, Acorus* and *Eleocharis* are ideal.

Tip 212 *A paludarium focuses on decor and plants.*

213 How do I water terrestrial plants in a paludarium?

The outlet of an external filter is a good water delivery system for paludariums. Spraybar extension kits are available to distribute water across the entire length of the aquarium. Direct the flow onto the plant roots or the material in which the plants are rooted. If you set it up properly, you will never need to water the terrestrial plants featured in the display.

214 Keep the air above plants cool in summer

Strong lighting can heat the air above the water, causing heat fluctuations during the warm summer months and burning the leaves of any emergent plants. To alleviate this problem, simply use a small fan directed toward the water surface. Make sure the fan is positioned so it will not fall into the aquarium. For closed hoods or small spaces, use a battery-powered fan and only use it when required.

215 Extend the display outside the aquarium

To bring a display out of the box, try placing suitable houseplants along with pieces of wood and rocks around the outside of the aquarium. For the best effect, use the same style of decor inside and outside the tank. In open-top aquariums, position some pieces part inside and part outside.

216 Clean away the algae, but leave some for the fish

Complicated displays can be difficult to clean, with lots of crevices and gaps where algae can grow. To clean these hard-to-reach areas, simply use a small toothbrush. However, many species of fish and shrimp are also great algae eaters and aquarium cleaners, so remember their needs when planning your display.

217 Clear away debris in a heavily planted aquarium

In a heavily aquascaped aquarium, especially one with many plants, it can be difficult to remove the build-up of debris on the aquarium floor without damaging the display. A good strategy is to place a few pieces of bogwood on slightly dipped or lower patches of substrate. Debris will then collect by gravity at the base of the wood, where you can easily siphon it away.

Tip 218 *Hide airstones and powerheads behind rocks.*

218 Bubbling rocks make an intriguing feature

For an unusual feature and focal point in the aquarium, place airstones under piles of rocks and hide the air line under the substrate and behind decor. Adjust the flow to suit your needs, and the air bubbles will mysteriously rise from the top of the rocks. The same effect can be achieved with hollow decor, such as artificial barnacles or suitable pieces of wood. Make sure the airstones are positioned to allow easy access for periodic replacement or cleaning.

Tip 219 *A concealed water pump generates white water.*

219 Swirling streams suit fast-moving fish

Use air pumps connected to airstones to create a sense of movement in fast-flowing stream- or river-style aquascapes. For the best effect, place the stream of bubbles in front of a water pump outlet. The water movement can then be seen agitating and swirling the bubbles around the aquarium. Along with suitably fast-moving and active fish, this effect will create a lively and realistic display.

220 Build in extra pump power, but make it removable

When using pumps to provide water movement around a rockscape, remember that you will need to remove any pumps for maintenance. Creating water movement in the right areas may require incorporating pipework into the rockscape. Fit one end of the pipe in the area where you want to create movement and leave the other end poking out of the rockscape toward the bottom. Attach the pump here and hide it under a removable rock or two, where you can reach it for maintenance.

221 Making one pump do the work of many

Rather than using several pumps to create a lot of water movement, use one large pump with a flow-controlling divider or T-piece. This will make the aquarium easier to maintain, reduce the space used by the equipment, and give greater control over flow rates. Dividers normally sold for water features can split a pipe into two, three or four outlets, each with an adjustable flow.

222 Simulate the effect of sunlight overhead

To recreate the appearance of sunlight through patches of overhead vegetation, use some carefully positioned spotlights to focus light onto specific areas of the tank. You can use this method to highlight features such as a piece of wood or a specimen plant. Just be careful if you have a small tank, as spotlights can get very hot and easily overheat smaller tanks.

223 How do I create a mist effect in my display tank?

For aquariums with a lowered water level, you can create a stunning effect with a mister or fogger unit. These tiny units produce an eerie mist, which floats on the water surface and looks particularly good in paludariums or heavily planted aquariums. Misters also have the effect of creating a humid environment, welcomed by plants above the waterline that may otherwise suffer from the heat given out by strong lighting.

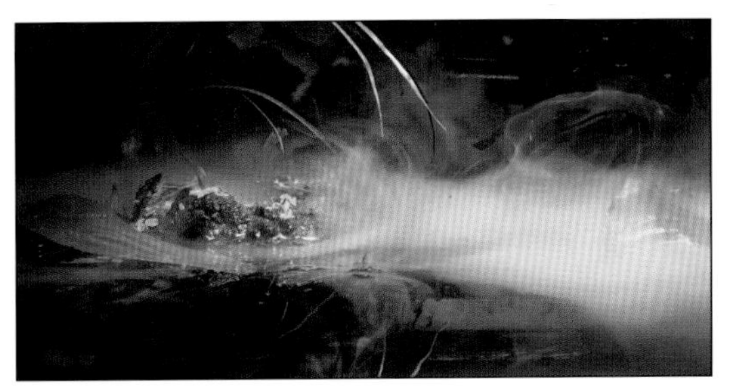

Tip 223 *Misters create humidity.*

224 Keep looking for new ideas and sources of decor

Finding a reliable retailer is a vital part of the fishkeeper's armory. However, while dealers can offer advice and provide good-quality livestock, they may not have the variety of decor found elsewhere. Look around aquatic stores, stone merchants and garden centers for ideas and inspiration. Many good displays are built around an unusual piece of decor.

Better Fish Buying

225 Find out all you can about a fish before buying it

Never buy a fish on impulse. First ensure that you are not taking on something that will cause you or its tankmates problems now or in the future. Possible pitfalls could relate to compatibility, diet, size or water chemistry that you cannot easily replicate in your own aquarium. If you are unsure about the suitability of a fish seek your dealer's advice. If your dealer can't help, ask to have the fish reserved for you until you can do more research.

Tip 230 *Look carefully at all the fish in the display tanks.*

226 Build up your fishkeeping skills and confidence

If you are new to fishkeeping, start with fish with a proven track record: easily-maintained, farm-bred community species, such as small barbs, danios and tetras. These will build your confidence to try something more challenging as you gain experience, whereas early failures can be discouraging and lead you to abandoning the hobby. Most specialist fishkeepers learned the basic skills with a community tank before moving forward.

227 Try to source your fish close to home

If possible, buy fish from a local dealer. Their water chemistry is likely to be close to your own, and therefore the community fish for sale will adapt readily to your (conditioned) tap water. Always ask whether the water in a sales tank has been specially prepared, for example, to meet the needs of soft-water species.

228 Growing fish will need more space in the aquarium

It is most likely that the fish you see in stores will be juveniles rather than adults. This makes commercial sense for your dealer, as they can buy them relatively cheaply and offer more individuals in each sales tank. However, you must find out the fish's maximum adult size and make due allowance for growth. Base your stocking levels on the size your fish are likely to attain, not their size at the time of purchase.

229 Pick a quiet time to go shopping

Try to visit aquatic stores midweek, when business is fairly quiet and staff have more time to discuss your needs. On a busy weekend, the retailer may advise you to return when he or she can spend more time guiding you through the choice of fish on offer. It is in the retailer's interest to look after customers, whatever their level of knowledge.

Tip 228 *Most fish on display are juveniles.*

Tip 233 *Balloon mollies* (Poecilia sphenops) *are no improvement on nature.*

230 Look at all the fish in the tank before you choose

Never buy fish from a sales tank where even one individual appears sick. There is a strong possibility that whatever the ailing fish is suffering from will be passed on to the others, even though outward symptoms may not manifest themselves until long after you have brought your fish home. It's also a reflection on the care (or lack of care) your dealer takes with their stock. (See also Tip 453.)

231 Can I ask a dealer to select particular fish from a tank?

It's reasonable to specify individual fish if you want a pair, or if some are obviously larger or showing better color than others of the same species. But don't expect staff to net, say, individual neon tetras *(Paracheirodon innesi)* from a tank that may contain hundreds of virtually identical specimens—it's not only impossible, but stressful for the fish.

232 Quote the scientific name when ordering fish

If your local store does not have the fish you want, it can usually order them for you if you don't mind waiting for the next consignment from the wholesaler. To be certain you are getting the right fish, quote the scientific (Latin), not the common (English) name—a term such as "zebra fish" can lead to confusion, as it can apply to anything from cichlids to danios to catfish.

233 "Man-made" fish may be at a disadvantage

Some tropical fish are selectively bred to display traits that do not necessarily work in favor of their health and well-being. An example is the so-called balloon molly *(Poecilia sphenops)*, which has a rounded, shortened body. This makes it prone to swim-bladder problems, as the organ is severely compressed. Ask yourself whether it is ethical to buy such fish when there is so much natural beauty to choose from.

234 Points to look out for when buying fish

Healthy fish deport well, swimming in a way typical of their species. The eyes are clear, neither protuberant nor sunken; dorsal fins are erect and other finnage is carried clear of the body, with no serious splits, reddening or fraying. Missing or raised scales should ring warning bells. Fish passing this visual check may have diseases yet to show themselves, but it makes sense to start with specimens manifesting no outward health problems.

235 How can I be sure of buying fish of both sexes?

If you are buying immature school fish with no obvious external sexual differences, and you want males and females with a view to spawning them later, a group of half a dozen will give you a better than 90 percent chance of obtaining both sexes. Female egg-layers also tend to be deeper-bodied than males, even before they are roeing up. (See also Tips 379 and 380.)

Female

Male

White Cloud Mountain minnows (Tanichthys albonubes)

Better Fish Buying

236 Should I buy all my breeding stock from the same outlet?

When buying breeding stock, try to purchase males and females from different stores. This makes it less likely that the fish are from the same brood, perhaps bought in from a local hobbyist, and offspring will be less likely to suffer the effects of inbreeding. Be especially wary of species that easily reproduce and throw large broods, as these most often find their way from hobbyists into the stores.

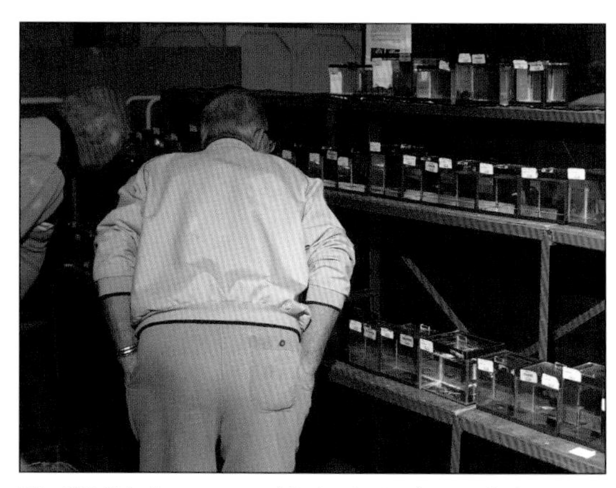

Tip 239 *Fish shows are good for inspiration, but not for buying.*

237 Is it a good idea to buy wild-caught fish?

Wild-caught fish are tempting buys, usually being larger, more vibrant in color and truer to type than tank-breds. However, they are also more expensive and potentially harder to keep, as you will need to match your water parameters closely to those of their habitat. A better purchase for the beginner would be F.1.s (first generation tank-bred fish).

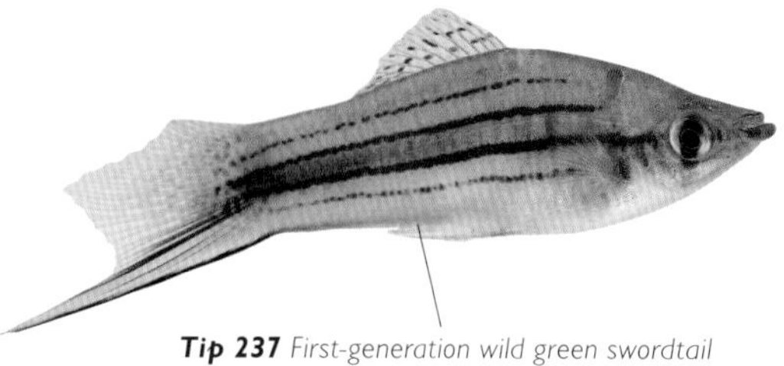

Tip 237 *First-generation wild green swordtail* (Xiphophorus helleri) *cross.*

238 Does it matter if the color of fish on sale seems muted?

The colors of fish you see for sale may appear washed out compared to photos in books and magazines. This is not necessarily cause for concern: for convenience, many sales tanks are brightly lit, sparsely furnished, with no substrate. Fish show their best coloration against a darkish substrate, and when refuges are provided.

239 Is it safe to buy fish on display at shows?

However tempting it may be, don't buy fish from shows. They will be stressed by the move from aquatic store to another destination, and the water available at shows is not always of the correct chemical composition, and it may not even be dechlorinated. Add to this the inevitable tapping on the tank glass by passing members of the public and the odds are stacked against the fish surviving.

Tip 238 *Young rainbowfish which have yet to attain their full color.*

How fish are bagged up

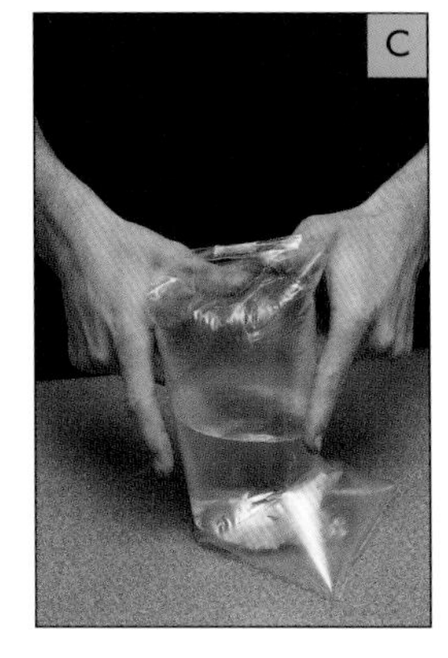

A *Once the dealer has placed your fish in a plastic bag, take the opportunity to view them from below to check they are healthy.*

B *When you are satisfied, the dealer will tie a tight knot in the top of the bag, trapping air and water for the fish's journey home.*

C *To prevent the fish being trapped in the corners, this dealer slides the upturned bag into a second bag, tucking in the corners.*

D *The dealer then places both bags into a green bag so that the fish travel in the dark to reduce stress. He adds care instructions.*

240 Ask to examine the fish in their bag before buying

Once your purchase is netted, but before the bag is inflated with oxygen and knotted or secured with an elastic band, hold it up and examine the underside of the fish for signs of damage or infection. You may spot something amiss that was not apparent when the fish was swimming in the tank—hidden faults are especially likely with bottom-dwelling species.

241 Buying the newest imports will be expensive

If you are on a tight budget, and new fish come into the hobby that you like the look of, it will often pay to wait a year or two before buying as their price will likely come down. This will happen if the fish prove a commercial success and are able to be farm-bred. The first imported neons were hugely expensive compared to their low price today.

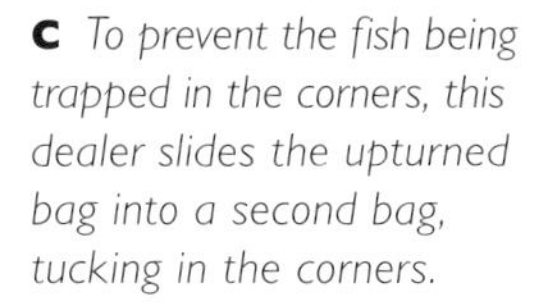

242 Are tiny fish at greater risk during transport?

When you are purchasing very small fish, ask for the corners of their plastic transit bag to be folded over and taped up. This prevents the fish from becoming trapped, and possibly crushed, and it also lessens the likelihood of any being accidentally left behind when the fish are transferred to your aquarium.

243 Is it safe to transport spiny fish in a plastic bag?

If you are buying fish with sharp fin spines, such as armoured catfish, or teeth (like piranha), bring along a white plastic bucket with a securely fitting lid to take them home. Plastic bags are certain to be punctured in transit, even if the fish are double-bagged. Failing a bucket, a Styrofoam fish box will do, although when partially filled with water it will be heavy to carry. You may need help with transportation.

Better Fish Buying

244 Using a Styrofoam box to get your fish home

Once purchased, get your fish to their destination without delay. Place bags in a lidded Styrofoam fish box, which provides insulation and shielding from stressful light. Pack crumpled newspaper or empty inflated plastic bags between the fish bags to prevent them rolling around, then put the box in your car trunk or on the backseat floor and drive home slowly, with no unnecessary stops along the way.

Tip 244 *A lidded box is useful for transportation.*

245 Transferring fish from a bag to the aquarium

To transfer new fish to an aquarium or quarantine tank, open the bag, roll it down to form a collar, and float it in the tank (with the lights off). Leave just long enough for temperatures inside and outside the bag to equalize, then tilt it so the occupants can swim out. Scooping tank water into the bag to acclimatize fish is pointless and needlessly stressful. They take weeks or months to adjust fully to changes in pH and other parameters.

246 Can I pour the water from a transit bag into my tank?

Some fishkeepers will not tip transit water into their aquarium because it is ammonia-laden. However, unless you are unloading several large bags into a small tank, dilution and filtration will take care of any pollutants. The main concern is not to stress the fish, which you will do if you lift them out of their bags.

247 My new fish look healthy. Do I need to quarantine them?

You should always quarantine new purchases before adding them to an existing community. The tank need only be basic, with heaterstat, air-driven sponge filter and a refuge for the fish. A week or two in solitary is good insurance against introducing stress-related diseases such as ich to your main aquarium. And if a new fish does fall sick, it can be medicated in situ, avoiding possible adverse side effects to your existing stock.

Tip 245 *Roll down the sides of the bag and float it in the tank.*

248 How will my existing fish react to newcomers?

It helps to rearrange the decor in your tank before introducing new fish. That way, any territorial occupants, such as red-tailed black sharks, will be preoccupied with the changes and less likely to bully the newcomers before they have a chance to settle in.

249 My new fish aren't feeding. What should I do?

Don't worry if new introductions take a while to feed: they will be stressed by the move, and the best thing you can do is leave them in peace for a day or so with the lights out. If they are healthy specimens, missing a meal or two will do them no harm.

250 Will a dealer replace any fish that die?

Ask the dealer whether fish you buy are guaranteed. A few stores operate an unconditional replacement policy on livestock that dies within a specified period (a few days at most). However, the majority either do not guarantee their fish or insist that any casualties are brought back with a sample of your aquarium water for analysis, to establish that it was not your poor husbandry that killed them.

251 I've just set up a tank. Can I add the fish in any order?

The first occupants to buy for a planted community tank are small and inexpensive algae-eaters, such as *Otocinclus affinis*. It is best to leave the largest fish for last, once the system is fully mature, otherwise they can place a strain on the filter. The list on the right reflects a selection of fish that could be added at different stages as the system matures. Always check on the water requirements and compatibility of fish you wish to add to a community aquarium.

Adding fish

First fish

- Zebra, leopard and pearl danios
- Black widow, lemon and silver-tipped tetras
- White Cloud Mountain minnow
- Cherry and golden barbs

Second fish

- Arulius, black ruby, black-spot, blue-barred, chequered, clown, Cumming's, odessa, rosy and tiger barbs
- Clown, red-striped and scissortail rasboras
- Black neon, bleeding heart, Buenos Aires, cardinal, Congo, emperor, flame, glowlight, head-and-tail light, neon and serpae tetras
- Croaking, dwarf, honey, moonlight, pearl, snakeskin, thick-lipped and three-spot gouramis
- Corydoras catfish as available, such as bronze, dwarf, Harald Schultz's, panda, peppered and Sterba's
- Clown and dwarf loaches
- Guppies, platies, sailfin mollies and swordtails

Third fish

- Black phantom, red phantom and rummy nose tetras
- Red-breasted flag cichlid and *Nanochromis parulis*
- Orange aggie, Borelli's dwarf, chequerboard, cockatoo dwarf, keyhole, Nijsseni's dwarf, striped dwarf and viejita dwarf cichlids
- Kribensis plec and ram cichlids
- Bristlenose catfish and clown or zebra plecs
- Glass, twig and whiptail catfish
- Hong Kong and kuhli loaches
- Banded, red, threadfin, boesmani, splendid and dwarf rainbowfish
- One-lined and three-lined pencilfish
- Headstander and marbled hatchetfish
- Lyretail killifish
- Betta (only one male to a tank)

Better Feeding

252 How can I tell what sort of food my fish prefer?

If you buy a fish that is something of an unknown quantity, and you are not sure of its feeding habits, the position of its mouth will often give you valuable clues. Upturned mouths indicate surface-feeders (insect food); level mouths suggest midwater feeders (omnivores); and underslung mouths, often surrounded by barbels, are typical of bottom-feeders.

A longer lower jaw shows that the fish feeds by approaching food from beneath.

Terminal mouths are typical of midwater fish that approach their food head-on.

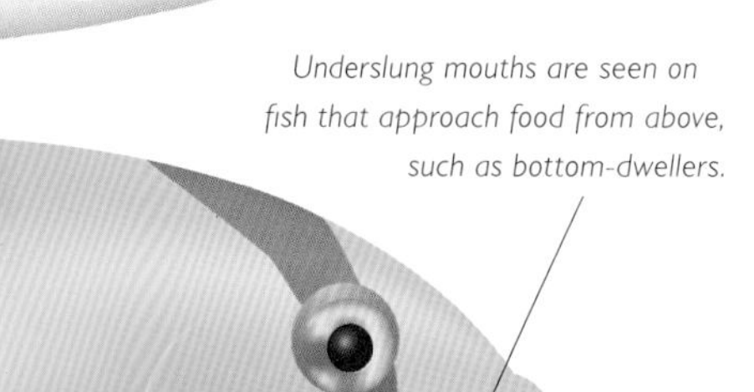

Tip 252 *A fish's mouth offers clues to its diet.*

253 Preserve the goodness in flaked fish food

The vitamins in flake food start to degrade from the moment the container is opened, and this process is sped up by exposure to heat, light and a damp atmosphere. Store tubs of flake in a cool, dry place—never on top of a radiator or on the tank hood—and replace the lid immediately after every feed.

254 Can I use the powder at the bottom of a tub of pellets?

Don't tip it into the tank but sieve it off, otherwise it just forms a film on the water that, as it decomposes, adds to the ammonia loading. If some powder does get through, skim it off immediately by drawing a sheet of plain, white paper towel over the water surface.

255 Sinking granules will get stuck in aquarium gravel

Sinking granular food is ideal for bottom-dwelling catfish such as corydoras and brochis. But don't feed it if the substrate is made up of coarse aquarium gravel: particles will only lodge in the crevices and decompose. On a sandy substrate, all the granules will be picked up and there is less chance of the fish damaging their barbels.

Tip 255 *Sinking foods may lodge in coarse gravel.*

256 Choose the right-sized food for the fish in your tank

If your tank contains fish of widely differing sizes, tailor the particle size of stick food and pellet foods to the mouths of the smallest, not the largest, inhabitants. Large-particle food is unlikely to soften in time for the small fish to get their share. Stick foods can be easily snapped into manageable portions, while pellets are available in all grades from fine to coarse.

Additives in food can enhance the colors of tropical fish.

Flake foods can provide the staple diet for most fish. Offer them sparingly.

Stick foods are useful for feeding the larger barbs.

Pellets are available in floating or sinking types for all tank regions.

257 Keep dried foods dry in their containers

Moisture in containers of flaked food encourages mold. The most common way this food gets damp is when you take a pinch, feed it to the fish, get your fingers wet and then dip them back into the tub. To stop this happening, always keep an old towel by the tank so you can dry off your hands.

258 Buying in bulk makes sense for a group of hobbyists

Large tubs of flake food work out to be more economical than small ones, but by the time you come to the end of yours it could well be past its expiry date. If you have friends in the hobby, it pays to buy the large sizes and split them up between you—that way, all the food is used up while it still retains maximum nutritional value.

259 Help! I've run out of flake food. What should I do?

If you run out of tropical flake, it is safe to feed cold-water flaked food for a day or two, but don't make a habit of it just because it tends to be slightly cheaper. The formulas cater to different nutritional needs and metabolisms, and long-term the two types of food are not interchangeable.

Tip 257 *Don't put your wet fingers in a tub of flake food.*

Better Feeding

260 Variety ensures the best diet for your fish

Try to offer a varied diet, even if your fish seem perfectly happy with good-quality flake food. The wider the menu, the more likely it is to provide all the essential vitamins, minerals and trace elements. A flake-only diet may encourage the condition known as Malawi Bloat in Rift Lake cichlids, and it may not be suitable for species that naturally consume a great deal of vegetable matter.

261 How can I be sure that all my fish are getting food?

When feeding flake food to a community of fish made up of surface, midwater and bottom-feeders, hold a pinch under the water for a few seconds before releasing it. Some of the flakes will then sink, making it more likely that all the fish will get their fair share. If all the food is left to float, there is always the risk that the surface-feeders, such as guppies, will monopolize the meal at the expense of the other fish.

Tip 261 *Make sure all your fish get a share of the meal.*

262 Add store-bought live foods to the aquarium

Store-bought aquatic live foods, such as bloodworm or *Daphnia,* come bagged in water. Don't just tip the bags into your tank, instead strain the contents through a fine-meshed nylon net and then invert it under the surface. Keep the net solely for this purpose, and sterilize it after use by hanging it in the sun to dry, which is safer than using disinfectant.

Tip 262 *Bloodworm is a welcome treat food.*

263 Can I offer my fish worms from the garden?

Earthworms dug from a chemical-free garden, or collected on damp nights from lawns, are a food few fish can resist, and they are protein-packed. Red worms from angling stores are a good alternative if you don't have a garden or yard. But take care not to feed them to the exclusion of other foods, otherwise fish can easily become fixated on them and refuse all other offerings.

Red worms sold for angling are an ideal earthworm substitute.

264 Small fish won't be able to eat whole earthworms

Earthworms should be chopped for feeding to your smaller fish. This is not a pleasant task, so the quicker it can be done the better. A great help is a pair of multi-bladed scissors, sold in fishing tackle stores to anglers for chopping up their worm baits. Alternatively, use a metal pastry-cutter (on a chopping board kept solely for this purpose).

265 How to make a feeding ring and floating worm feeder

Make your own feeding ring for floating food any size you like by taking a piece of air line and pushing one end into the other. Use silicone sealant on the join if you need to. To make a worm feeder, punch holes in a clear plastic tub (the sort single-portion desserts come in) and make an air line buoyancy collar to fit.

266 Can I feed my fish with insects from the garden?

There is no risk of introducing disease to the aquarium by feeding live insect foods gathered on land or dug up from the garden. They may carry pathogens, but these are specific to a terrestrial, not an aquatic, environment. Just avoid collecting brightly colored insects, as they are advertising the fact that they are unpalatable, or even toxic, to animals that might otherwise be tempted to eat them.

267 Is there such a thing as safe live food from a pond?

The safest aquatic live foods you can gather are from ponds that contain no fish—the crustaceans and other small creatures you catch cannot be intermediate hosts to fish parasites. Nutrient-rich cattle drink ponds are often alive with *Daphnia* blooms, but always remember to ask the owner's permission before collecting from such sources.

Tip 265
A homemade feeder with buoyancy collar (left).

Freeze-dried bloodworm will float for a time and stay within the air line ring (below).

Better Feeding

268 Live food from the pond filter to the aquarium

If you own a garden pond with a filter, you can gather useful live foods for your tropical fish. Inside the filter chambers you will usually discover aquatic hog lice *(Asellus)*, which look rather like wood lice, and feed on detritus. Net them out, rinse them in the net under a running tap and watch your fish devour them as they tumble through the water.

269 Are angler's maggots a safe live food for aquarium fish?

Maggots, if you can find them, are a cheap and nutritious live food. Try a bait store. Feed only the white (uncolored) type, as dyes used to make them a more attractive bait may contain harmful chemicals. For smaller fish, try squatts (the larvae of the housefly, sold in red foundry sand); just wash before use. Pinkies (greenbottle larvae) fall midway between the two in size.

Tip 269 *Uncolored maggots are a good live food.*

270 How do I know which frozen foods my fish will accept?

Gamma-irradiated frozen foods are often sold in variety blister packs containing, for example, bloodworms, *Daphnia* and glassworms. Buy one of these first and note which varieties your fish prefer. Afterward you can buy the favored foods in single-variety blister packs.

Tip 271 *One defrosted cube of bloodworms makes a hearty meal.*

271 Is it safe to feed tubifex worms to my fish?

Live tubifex worms live in heavily polluted water and should never be fed to fish. Even if your aquatic store claims to have scoured them under a dripping tap, they can still pass on disease. The freeze-dried cubes, however, are perfectly safe. Try sticking them to the tank glass at varying depths so that all your fish can benefit from them.

272 Will a special diet encourage my fish to spawn?

Fish often need live foods to bring them into spawning condition. One of the best is whiteworm, easily reared in a shallow wooden box filled with damp peat. Mix in a starter culture, follow with a slice of bread and top it off with a close-fitting sheet of glass. The worms feed on the bread and can be picked from the peat below with tweezers and fed directly to the fish. You may need to obtain a starter culture from a fellow hobbyist.

Tip 270 *Variety packs of frozen food offer your fish a choice.*

273 What food should I offer my surface-feeding fish?

Fish that spend all their time at the surface, such as butterflyfish *(Pantodon buchholzi)*, may refuse everything you offer them except live insect food. Spiders, woodlice, moths and beetles, or black crickets from stores that cater to reptile-keepers, are all ideal. Just make sure the tank has a tightly fitting lid—escaped crickets head for the smallest crevices and are virtually impossible to find and get rid of.

Tip 273 *The predatory butterflyfish seeks out aquatic insects and small fish. Keep it with tankmates too large for it to eat.*

274 Maintain a regular supply of whiteworms

Have several whiteworm boxes on the go simultaneously, so that when one starts to go stale you can re-seed it with fresh peat and young worms and still harvest a constant supply from the others. If the bread does not need replenishing every two days, you are taking too many worms from the box, and should rest it for a couple of weeks. Whiteworms are, however, very fatty and should be offered sparingly.

275 Where can I buy food for carnivorous fish?

Your fish store is a good source of food for all carnivorous species. Freshly boiled cockles, chopped raw mussels and scallops, prawns (left in their shells for large fish), shrimps, and boiled cod roe are all on the menu, but if they remain uneaten they will quickly pollute the tank. Long-handled aquarium tongs are handy for removing leftover morsels.

276 Do my piranha need a diet of live fish?

It is cruel and unnecessary to feed live fish to piranha *(Pygocentrus nattereri)*, which can be started off on earthworms when young and then weaned onto strips of oily fish (sprat, herring, trout) or thawed frozen lancefish. But if you regularly have access to large quantities of unsaleable fry, keeping a piscivore such as an oscar *(Astronotus ocellatus)* is an acceptable way of disposing of them—and keep the predator happy.

Tip 276 *This young piranha will accept earthworms or surplus fry.*

277 Which green foods can I offer my herbivorous fish?

Herbivorous fish such as plecs appreciate fresh greens. Suitable items include blanched lettuce leaves, fresh (uncooked) garden peas, pinched to remove the skins, and raw zucchini or cucumber slices. Non-toxic lettuce clips are available from aquatic stores to hold items down in the water, or you can sandwich leaves in place with an algae magnet. It may take time to acclimatize fish to green foods, but persevere.

Tip 277 *Secure lettuce leaves in a clip so that the fish can graze on them. Left to float, the fish usually ignore them. Remove uneaten food before it starts to rot.*

Blanched lettuce leaves

Fresh peas with their skins removed

Cucumber pieces and zucchini slices

278 Certain foods should never be offered to fish

Never feed your fish meat scraps, cheese or any items containing animal fat. Fish require fats, but only in the form of lipids (fatty acids), which do not solidify and damage the internal organs. Also avoid sugary foods (including those processed in sugar, such as baked beans), and white bread and cracker crumbs, as these contain saturated fats.

279 An acceptable homemade meat treat for fish

Beef heart and ox liver are exceptions to the "no feeding meat to fish" rule, as both are very low in fat. Chop them, stir in quality flake and a little melted gelatin as a binder, and then spoon the mush into an ice cube tray and freeze. Pop out individual portions to feed. Discus are especially fond of beef heart, but offer it only sparingly.

280 Consider buying a freezer to store frozen fish foods

If you regularly feed frozen foods (which are just as nutritious as the live items), invest in a small freezer of your own in which to store them. As well as avoiding domestic conflict, this will ensure you can find what you want, when you want it, rather than rummaging through the contents of the family's groceries.

281 Should I discard frozen foods after a power failure?

If your freezer or the power fails, treat frozen fish foods just as you would items for human consumption. If you discover the fault before the packs have defrosted, it is safe to move them to another freezer; if they have partially thawed, feed normal rations on the day and discard the rest; if they have thawed out completely, throw them away immediately. Never refreeze fish foods once defrosted.

282 How should I store foods that I have frozen myself?

Don't discard the blister packs that hold frozen foods. They are good alternatives to ice cube trays when making up your own recipes, take up less space in the freezer, and portions are a more manageable size. To feed even smaller portions, score across each helping of food with a knife before freezing, and it should snap in two without much trouble. Wash out the packs thoroughly before reusing them and label them clearly with their new contents.

283 Keep the damp out of opened food packs

Freeze-dried foods have all their moisture content removed but retain most of their natural goodness. To keep the contents of larger tubs bone-dry, tape a sachet of silica gel granules (as sold with cameras, clocks and electronic equipment) to the underside of the lid.

Silica gel keeps freeze-dried foods dry once the packs are open.

Tip 284 *Brine shrimp are suitable for freshwater fish.*

284 Is it safe to put brine shrimp into a freshwater tank?

Adult brine shrimp *(Artemia)* are a fine treat food. However, they are supplied in bags of salt water, which you don't want going into your tank. Tip the contents into a bowl and suck up the shrimps with a meat baster. By varying the depth at which you release them into the tank, you can dictate which type of fish get to them first—surface, midwater or bottom-feeders.

285 What is the best food for fish fry?

Newly hatched egg-layers will need much finer fare than powdered flake. Liquid fry food (minute particles in suspension) will see them through the first few days, after which most species can be weaned onto brine shrimp nauplii. If you are expecting a brood, start your culture(s) well in advance—don't wait until the fry are free swimming in the aquarium, as without food they will die within hours. (See also Tips 400 to 407.)

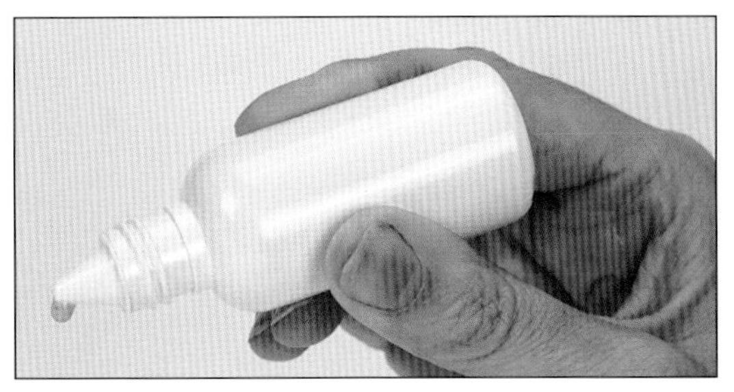

Tip 285 *Liquid fry food is commercially available.*

286 Feed fish fry on the basis of little and often

Unlike adult fish, fry need small, regular meals throughout the day if they are to grow properly. Automatic food dispensers, clockwork or electronic, filled with powder-fine granular foods are useful in fry tanks, but don't rely on them for long periods of absence. Granules can easily become damp in the humidity above the tank and clump together, so the feeder no longer discharges as it should.

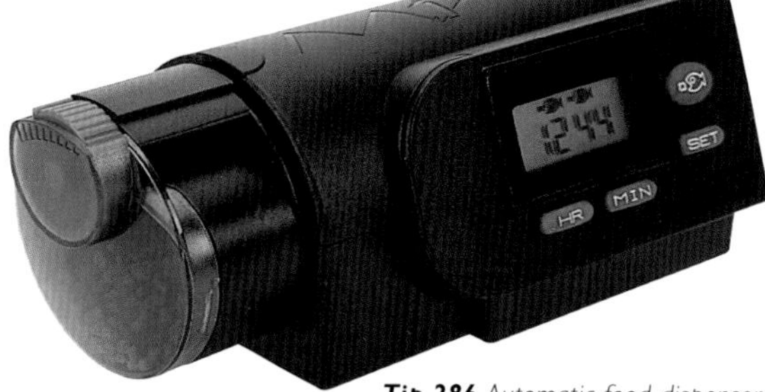

Tip 286 *Automatic food dispenser.*

Tip 288 *Pacu* (Colossoma macropomum) *go crazy for hazelnuts.*

287 Air-powered sponge filters are a source of free fry food

Air-powered sponge filters are ideal for fry tanks—not only are the young fish in no danger of being sucked into them, but the surface of the sponge becomes colonized with microorganisms in fry tanks such as rotifers, which the babies will graze upon. Two filters per tank are ideal, so they can be rinsed in rotation; that way, one or the other will always be a source of the bonus food.

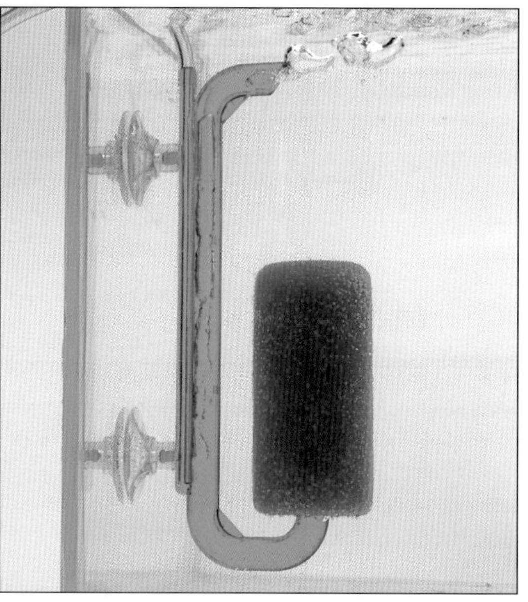

Tip 287 *Air-powered sponge filters are ideal for fry tanks.*

288 Fruit- and nut-eating pacu have strong teeth

Pacu *(Colossoma macropomum)* are commonly known as fruit-eating piranha and can be fed accordingly. They will take grapes, slices of banana and melon, and orange segments, but a real treat is hazelnuts in their shells. The fish wedge them against the tank glass and use their strong teeth to crack them open. There is a strong argument for using $^3/_4$-inch (20 mm) plate glazing on a pacu aquarium!

289 Pygmy puffers will help to control snail numbers

Freshwater pufferfish have rabbit-like teeth that continue to grow throughout their life. Snails form part of their natural diet, so if your tank is infested with the conical-shelled Malaysian live-bearing snails, which live in the gravel substrate, try introducing some pygmy puffers. They will keep snail numbers down, and chewing on the shells will stop their teeth from growing too long.

***Tip 289** A pufferfish investigates a snail in the aquarium.*

290 Should I turn off the filter while I feed my fish?

Get into the habit of turning off internal power filters for the duration of a feeding session, otherwise flakes will be sucked inside before the fish can reach them. On the other hand, finicky feeders will take fine granular food more readily if it is introduced in the outflow of the filter, as the movement this imparts fools them into thinking it is live food.

291 Check that fish are feeding before you buy them

Before committing yourself to buying expensive wild-caught fish, ask to see them being fed in the store after they have had a chance to settle. Newly imported fish may refuse food, and the longer the fast goes on, the more difficult it will be to persuade them to resume feeding. As a general rule, hungry fish are healthy fish.

292 How much and how often should I feed my fish?

More fishkeepers overfeed than underfeed. Fish will constantly solicit more than they need, and it can be hard to resist when they gather expectantly at the front of the tank. Limit feeds to as much as they will consume every five minutes, twice a day, and siphon off any remaining food after that time. Food left in the aquarium will only decay and produce ammonia, which can overload the filter.

293 Overfeeding attracts unwanted tankmates

Overfeeding may not lead to poor water quality if your filtration is good, but you can usually tell if you are overdoing it because the tank will be colonized by creatures other than fish, which capitalize on the surplus. These include planarian flatworms and ostracods—both harmless but a sign to cut down on the rations.

294 What is the best way to feed nocturnal catfish?

Nocturnal catfish can be secretive during the day, hiding away and missing out when you feed the other tank occupants. To make sure they get their rations, offer them sinking foods (pellets, algae wafers) just before lights out. However, turning the tank lighting off is no use if the room is still brightly lit. Feeding nocturnal fish should coincide with your own bedtime.

***Tip 294** Feed nocturnal catfish sinking algae wafers.*

295 Feeding aquarium fish when you go on holiday

If well fed before you leave, adult fish will come to no harm for up to two weeks while you go on holiday, and they will arguably be better for their fast. For longer periods away, make up daily rations of flake in twists of foil for a friend or neighbor to feed. Hide the tub so that he or she can't give in to the urge to give the fish that little bit extra.

Tip 295 *Package daily rations in twists of foil.*

296 Fish have broader feeding tastes than you think

Although some tropical fish are classed as herbivores and others as carnivores, all but out-and-out predators have a mixed diet in the wild. Algae-eating bristlenose plecs *(Ancistrus dolichopterus)*, for example, also eat the tiny creatures found amid the greenery, while livebearers take insect food as well as soft plant matter. So don't pigeonhole your fish when it comes to food—they probably have more varied tastes than you imagine.

297 Why do my plecs appear to eat bogwood?

The loricariid catfish (*Loricaridae* spp.), commonly known as plecs, require lignin to help digest their food properly. This is a constituent of bogwood, so always include an unvarnished piece in your decor if you keep fish of this type. They can then rasp away at it on demand. The much harder Mopani wood is no substitute for this.

Tip 297 *Bristlenose catfish* (A. temminckii) *will rasp away at bogwood.*

298 Garlic is a valuable food supplement for fish

Garlic is a powerful appetite stimulant and natural antibiotic that few fish can resist. To feed it, dampen some pelleted food, roll it in garlic powder and then freeze overnight so the flavor is drawn inside. Don't confuse garlic salt with garlic powder, and treat only a few pellets at a time so they can all be used up on the day they come out of the freezer.

299 Does it matter where I buy my fish food?

Unbranded bagged-up fish food may seem to offer good value for money, but you have no guarantee of its quality or freshness, and no redress if anything goes wrong. As in all things, you pay a bit more for quality, but bear in mind that fish are still very inexpensive pets to feed compared to warm-blooded animals.

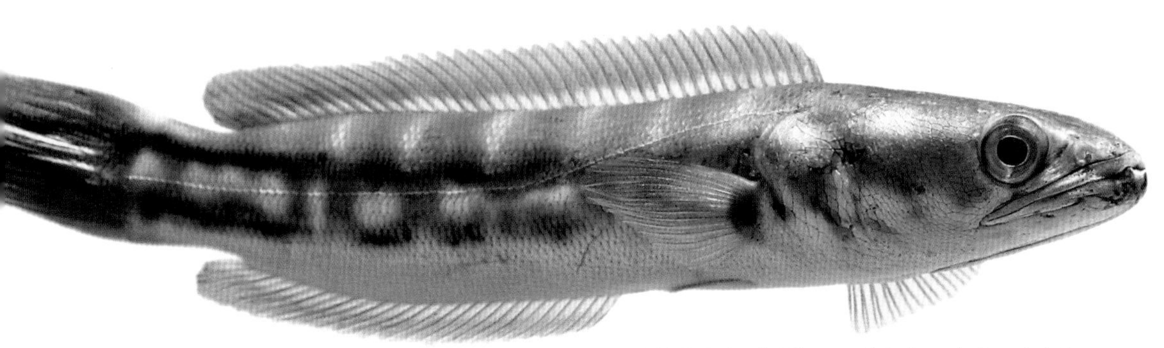

Tip 303 *Snakeheads* (Channa asiatica) *are dedicated piscivores.*

300 Should I worry if some fish are missing at feeding time?

Feeding time is an opportunity to check that your tank occupants are in good health, as you can usually count them as they come up to take food. But don't assume something is wrong if one or two are missing, especially those that brood their young, such as cave-dwelling cichlids or bristlenose catfish. They may have more pressing matters to attend to.

301 Don't allow messy feeders to pollute the water

Larger fish can be messy feeders when given pelleted food; they crush it in their throat teeth and expel fine particles through the gill covers. In such instances, it is best to time partial water changes to take place within an hour of a feed to minimize the polluting effects of the meal.

302 Another fresh treat for aquarium fish

All fish are cannibalistic to some degree, so a food that meets with universal approval is fresh trout roe (the eggs, not the milt from the males). Leave the membrane intact and it will be a hefty mouthful for large fish, or break it and crumble the contents so that smaller fish can enjoy the coral-colored individual eggs.

303 A satisfied fish won't view your fingers as food!

If you own aggressive fish, such as snakeheads *(Channa asiatica)* or some of the larger bagrid catfish (*Bagridae* spp.), and need to put your hands in their tanks for maintenance, feed them first! A filled-up fish is less likely to see your fingers as a legitimate appetizer and more likely to amble over to a corner to digest its meal.

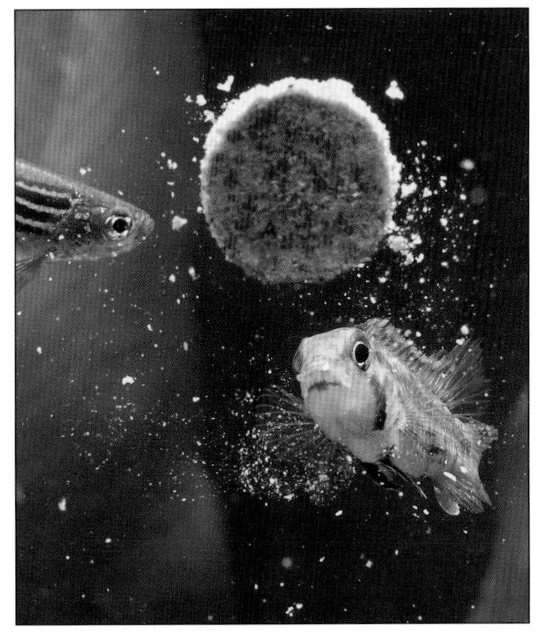

Tip 301 *Even tablet foods create a mess as they break up.*

304 Get to know your local aquatic dealer

Strike up an amiable relationship with your nearest good aquatic store. If you give them regular business, they will go the extra mile for you—ordering fish they don't immediately have in stock, or perhaps exchanging fish you have bred for dry goods such as food and equipment, so that your hobby becomes partly self-financing. (See also Tip 225.)

305 Do not disturb the aquarium unnecessarily

If everything is right with your tank, don't tinker with it in the intervals between the required regular routine maintenance. This will cause unnecessary disturbance to your fish, and may prevent them from breeding successfully.

306 Use your eyes as an early warning system

Learn what is normal behavior for your various fish, so that if they deviate from it you'll get a warning that something may be wrong. Your eyes are the best diagnostic tool at your disposal, and with self-training you can detect the early signs of water quality, compatibility and disease problems before they become acute. (See also Tip 456.)

307 Gain inspiration from public aquariums

Visit public aquariums. They have mature specimens of unusual fish you may be tempted to buy, and you can observe their behavior and decide whether they are really for you. Such places are also a source of ideas for aquascaping, particularly the biotope exhibits, where all plants and fish are from the same locality in the wild.

Tip 305 *Once established, disturb the aquarium as little as possible.*

Tip 308 *Algae can be beneficial in the aquarium.*

308 Algae are not always the aquarist's enemy

Algae are inevitable anywhere there is water, light and nutrients, such as in the aquarium. Instead of battling algae, consider that they make rockwork look natural, provide herbivore food, harbor microorganisms for fry to eat and remove nitrate from the water. In fact, algae are a useful nitrate indicator—the more algae, the more nitrate in your water. Clean the front glass, but otherwise leave algae alone.

309 Net cave-dwellers without stressing them

Chasing a cichlid or other cave-dwelling fish around a rock-filled tank with a net is usually a waste of time, and stressful for both fish and fishkeeper. Mouth-brooding cichlids may eat or eject their brood. Instead, position a large net at right angles to the front glass, use your hand to coax the fish out of the rocks and guide it into the net.

310 Modify a net to fit the aquarium

Nets have straight handles, but many aquariums have internal rims. Hence you can't press the net flat against the front glass to trap a fish. Buy nets with wire handles and bend them to create a right-angled step so the net fits flush to the aquarium glass. Use two nets: a bent-handle one near the front glass and a straight-handle one to chase the fish into this waiting trap.

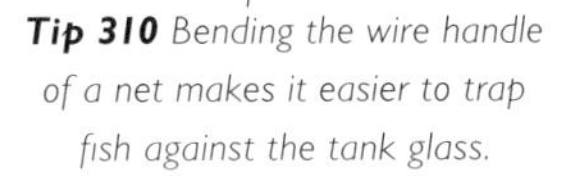

Tip 310 *Bending the wire handle of a net makes it easier to trap fish against the tank glass.*

311 How should I clean the inside front glass?

The best way to clean the inside front glass is with a dish sponge—the sort with a (usually green) plastic scourer on one side—reserved for aquarium use only. The outside can be cleaned with a vinegar solution (1 part vinegar, to 3 parts water), which is far safer than chemical glass cleaners (and their fumes) so near the tank water.

312 Avoid buying dyed and highly modified fish

Don't encourage the trade in dyed or genetically modified fish by buying them. The dyeing practice is cruel, unnecessary, and shortens the lives of fish subjected to it. Fish such as the blood red parrot and flowerhorn cichlids are engineered hybrids with uncertain temperaments and inherent deformities. (See also Tip 233.)

313 Touch test to check whether the heater is working

Get into the habit of touching the tank glass each time you pass by. This will indicate to you right away if a heater has stuck on or off, whereas looking at the thermometer requires much more conscious effort, and you may sometimes neglect to do it.

Tip 314 *Stick suckers to the tank glass with silicone sealant.*

314 Help suckers to do their job

The old-fashioned rubber suckers for thermometers, heater holders, etc., used to perish and disintegrate, but at least they stuck to the glass. Modern plastic ones are usually reluctant to do so, and become hard after a while and won't stick at all. Simply use aquarium-grade silicone sealant to stick them into place. The silicone can be cut away with a sharp knife, if required.

315 Setting up a mini-tank in an emergency

If you need a small tank in a hurry (for hospital, quarantine or fry-rearing) but don't have a spare one, use a large plastic jar, such as those used to display candies. Soak off the label and stand it in your aquarium to fill. No heater is required, although aeration is generally needed. Large plastic jars with a screw top are excellent for transporting fish; they don't leak and are easier to open than bags.

316 Breeding your fish means being prepared

If you intend to breed your fish, think beyond the arrival of fry and prepare live food cultures, overspill accommodation and perhaps a recuperation tank for one or more parents well in advance. It is pointless and cruel to have a tank full of babies if you lack the means to feed them in the first few days of life. (See also Breeding methods of popular fish, page 96.)

317 Setting up a large tank in an emergency

If a tank bursts, the fish can be temporarily housed in a non-toxic plastic garbage can with a heater and internal power filter. Alternatively, use a clean Styrofoam fish box, preferably one strengthened with a cardboard outer, as a temporary aquarium. Place it on a flat, solid surface and fill it about two-thirds full with water. Make sure the heater doesn't touch the Styrofoam. A fish box can also be used to transport a large fish or those with sharp spines (e.g. many catfish) that puncture bags.

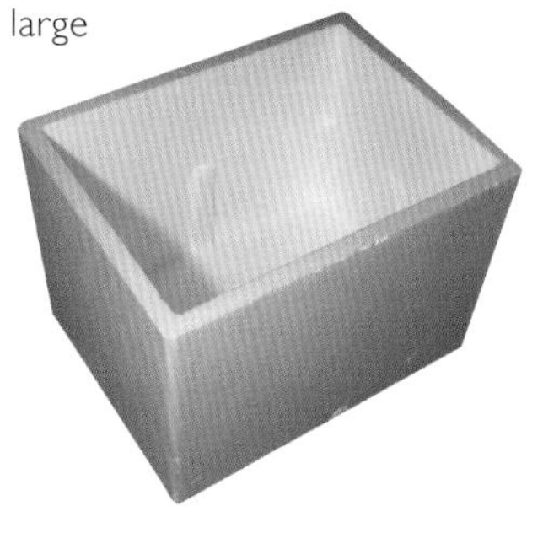

Tip 317 *A Styrofoam fish box has many uses.*

Tip 320 *Keep track of maintenance for a flourishing display.*

318 How do I protect my tank during a power failure?

If you have hot water available, fill plastic bags or soft drink bottles and suspend or stand them in the tank as emergency heaters. Cut down heat loss by insulating the tank with a blanket. Aeration and air-driven filtration can be operated with a bicycle or car tire pump—about five minutes pumping every half hour will be adequate. Alternatively, pump up a car inner tube and let it slowly bleed air into the system.

319 Keep careful records when breeding your fish

Do not give up on attempting to breed your fish, even after several failures. Record keeping, rather than guesswork, is the key to success. Note down water parameters and all other factors (lighting period, feeding regime, source and age of parent fish, etc.), so that if you change any of these it can be identified as that vital trigger if and when the fish do spawn.

320 A log book will keep track of regular maintenance

Don't trust your memory. Keep a log of when maintenance is due and tick off the tasks as they are performed. Break tasks down into daily, weekly, biweekly, monthly and occasional columns. Also record any incidences of disease, how you treated them and the outcome. It will build into a useful personal reference. (See also Routine maintenance, page 115.)

321 Environmental factors are at the root of many ills

Experts are agreed that at least 95 percent of illnesses in aquarium fish are caused by environmental factors rather than pathogens (parasites, fungi, bacteria). Also, many pathogenic diseases attack only when fish are weakened by stress, unsuitable water or poor diet. So unless your fish has an obvious pathogenic disease, such as ich (whitespot), don't instinctively reach for a medication, but look for an environmental cause. (See also Tip 460.)

Better Fishkeeping Skills

322 New does not always mean better

Technical innovation is no bad thing, but the fishkeeping marketplace is awash with products that have yet to prove their worth. Yet more cannot justify their price tag in terms of the benefits they confer on you or your fish. Don't automatically buy something just because it is new. Wait for others to evaluate it, then read the product reviews in aquatic magazines.

323 Save money but do not put your livestock at risk

Some ways of saving money in the hobby (such as collecting safe live foods or propagating your own aquatic plants) are acceptable, but don't cut corners if this jeopardizes the welfare of your fish. For example, second-hand electrical equipment is more prone to failure than new, and you won't know its history. If it comes down to a choice between buying a new fish or improving the living conditions of those you already own, always go for the latter option.

324 Seek help for yourself and encourage newcomers

Never be afraid to seek advice, either from fellow fishkeepers or aquatic store staff. It's far better to admit to a lack of knowledge than to try to bluff your way through. Likewise, give your time freely to beginners who require information, however obvious their questions may seem. We all had to start somewhere.

Tip 324 *Your dealer can help you learn about the hobby.*

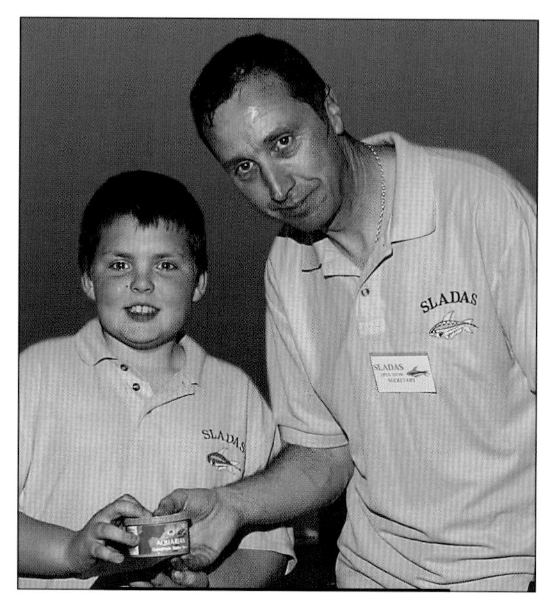

Tip 325 *A young fishkeeper receives a prize at a society fish show.*

325 Expand your knowledge by joining a society

Join a specialist fishkeeping society. Its members will have firsthand experience of the fish that interest you, and the information you can glean from them will be far more reliable than anything obtainable on the internet, where sound sense is inevitably mixed in with highly suspect advice.

326 Becoming a better aquarist is a nonstop learning process

Take every opportunity to learn more about fishkeeping techniques, water management and fish, and not just the ones you keep. Often some useless piece of information you have read can prove vital in dealing with an emergency. Or something you read about one fish can solve a problem with another. You can never know it all, but the more you learn, the better an aquarist you will become and the more you will enjoy it.

327 Scientific names are the key to accurate identification

Make sure you note the scientific (Latin) name of any fish you buy. Many common (English) names aren't universal, or may apply to more than one fish. If you need to look something up in a book or ask for advice, you will need the scientific name to make sure the information you obtain is for the right fish. Without the scientific name you may not be able to obtain any information at all! (See also Tip 232.)

328 You too can become a fish authority

Don't believe that breakthroughs in fishkeeping are the sole preserve of the experts. Close observation of what goes on in your tanks will sooner or later reveal fish behavior patterns nobody else has recorded. And just because a species has yet to be tank-bred does not mean it is impossible. Many ordinary hobbyists have achieved world firsts!

329 Help to conserve rare fish species

If you acquire species that are endangered in the wild, it is your duty to attempt to breed them. Specialist societies sometimes make such fish available to dedicated hobbyists. For example, the British Cichlid Association currently has several members boosting captive populations of rare African fish from Madagascar, Lake Victoria and Barombi Mbo, a crater lake in Cameroon, home to endangered cichlid species.

330 Pass on the fishkeeping bug to the next generation

Never forget why you entered fishkeeping in the first place: to enjoy the aesthetically pleasing living picture that is a community tank. Even if you move on to more specialized branches of the hobby, keep that first aquarium running in your living room so that the whole family can enjoy it. Your children may be inspired to follow your example.

Tip 330 *Colorful odessa barbs* (Puntius ticto *var.* odessa) *are a perfect living picture.*

331 Neon tetras thrive in conditioned tap water

Neon tetras *(Paracheirodon innesi)* thrive much more readily than cardinals *(P. axelrodi)* in conditioned tap water, even if it is hard and alkaline. This is because they are farm-bred, whereas most cardinals are still wild-caught from habitats where the water is soft and acidic, and they will not easily adapt to different parameters.

332 The elephantnose needs mature water

The elephantnose *(Gnathonemus petersi)*, with its trunklike snout, is often bought for its oddity value and is a very peaceful community species. However, it should not go into new systems as it requires mature water. It also needs the company of its own kind (keep it in groups of three or more), otherwise it will pine. Elephantnoses prefer shaded tanks with a leaf-litter substrate to root around in.

The marbled elephantnose (Camphylomormyrus tamandua) *uses its nerve-rich "trunk" to locate small live food items in the substrate.*

Tip 332 *Elephantnoses* (Gnathonemus petersi) *are shy creatures.*

Tip 333 *Neon tetras* (P. innesi) *make a display on their own.*

333 Big schools of neons are best kept on their own

The neon tetra *(Paracheirodon innesi)* is deservedly popular, but its small size restricts what can be kept with it—even traditionally peaceful species, such as larger barbs and tetras, will make a meal of it. The best way to appreciate neons and the closely related cardinals *(P. axelrodi)* is to devote a species tank to them, with schools numbering dozens or hundreds. It won't matter if you start with young specimens as nothing can threaten them.

334 Catching a spiny fish without damaging it

Catching a large plec (*Peckoltia* spp.) or doradid catfish (*Doradidae* spp.) from an aquarium can be difficult, as its hard fin rays are easily damaged in a net. A better method is to leave a suitably sized plastic container overnight in the tank, with a lid of plastic wrap in which you have cut a hole just large enough for the catfish to gain entry. Bait the trap with tablet food and the fish should be ready for removal by morning.

335 How can I get rid of the snails in my aquarium?

The best way of controlling snails is to keep fish that eat them. Clown loaches *(Botia macracantha)* are favorites, but if they are overfed they will not waste time on snails when easier pickings are to be had. Try cutting back on their rations. Some small South American doradid catfish (*Doradidae* spp.) are also good snail controllers; *Amblydoras hancocki*, Striped Raphael catfish (*Platydoras costatus)* and talking catfish *(Agamyxis pectinifrons)* are all good community species.

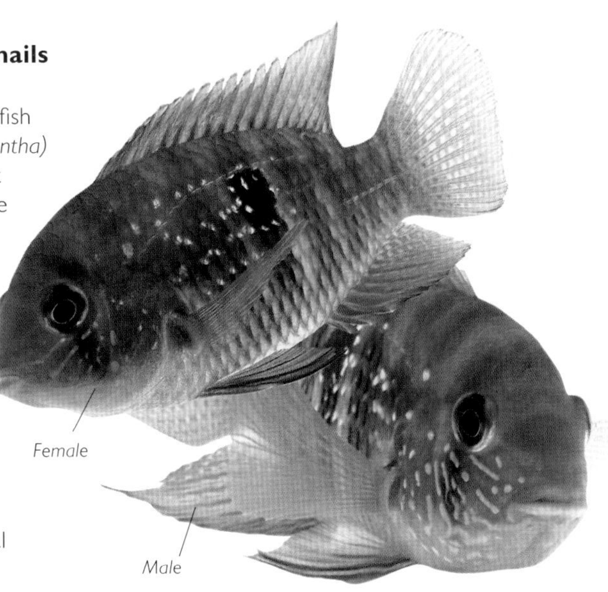

Tip 337 *Blue-point acara* ("Aequidens" coeruleopunctatus).

336 Guppies may be prone to guppy disease

Guppies *(Poecilia reticulata)* are not the ideal beginners' fish many people imagine them to be. Imports from Asia are weakened by hormone-induced breeding and are prone to guppy disease, thought to be caused by a tiny protozoan parasite of the skin and internal organs. Israeli- and Sri Lankan-bred guppies do not suffer from this, so inquire about the source of the stock you buy.

337 Provide refuges for female cichlids

Most male Central American cichlids are one-third larger than females, so to prevent bullying before a pair-bond has fully formed, place one or two clay drainage pipes into the tank, greater than the total length of the female but sufficiently narrow to prevent the male from entering. This will give her the refuge she needs if things turn rough.

Avoid keeping the golden sunrise (Poecilia reticulata) *with boisterous barbs or male bettas* (Betta splendens) *of the same coloration.*

A neon blue guppy (Poecilia reticulata). *The various strains may breed true if kept apart in the aquarium.*

338 Choose the best male swordtails for breeding

Some male swordtails *(Xiphophorus helleri)* attain sexual maturity very early in life, developing the full swordlike extension to the lower lobe of the caudal fin. These runts never attain the size of later-developing males, so when choosing breeding pairs go for males with shorter swords, as they will still have some growing to do.

Male pineapple swordtail (Xiphophorus helleri)

Male green-striped swordtail

Female green-striped swordtail (X. helleri)

Tip 340 *Variable platies* (Xiphophorus variatus) *need cooler conditions.*

340 Are all platies suitable for the tropical aquarium?

Two platy species are commonly available, but only the southern platy *(Xiphophorus maculatus)* is suited to tropical communities. The variable platy *(X. variatus)*, also known as the sunset platy, comes from eastern Mexico and thrives at between 64°F and 70°F (18°C and 21°C), far too low a temperature to be deemed a true tropical fish. To complicate matters, the species are interbred to produce color varieties that are tolerant of a wider temperature range.

339 Harlequins are at their best in large schools

Harlequins *(Rasbora heteromorpha)* are small, peaceful and colorful, but because they prefer the company of their own kind they do not make ideal community fish. They require soft, acidic water and broad-leaved plants to spawn successfully. Two-year-old males usually pair with females half their age. Younger males often show rudimentary ovaries!

341 A little salt in the water will help mollies

Black mollies *(Poecilia sphenops)* and green or golden sailfin mollies *(P. velifera)* are often sold as freshwater tropicals, when in reality they prefer their water brackish. Only with the addition of a little aquarium salt to their tank will they show their best colors and be less prone to shimmying (weaving from side to side while remaining in one spot).

342 Persuading insect-eaters to take flake food

Some fish, such as the butterflyfish *(Pantodon buchholzi)*, are difficult to wean onto flaked food because their diet in the wild consists of smaller fish or insects that fall into the water. An internal power filter, set with its return flow pipe positioned just beneath the surface, imparts movement to floating food. The butterflyfish then sees it as escaping prey and snaps it up. (See also Tip 273.)

343 Young discus in small groups are open to bullying

Don't buy young discus measuring 4 inches (10 cm) or less in groups of fewer than eight. If you do, the smallest or weakest fish will probably be bullied by the dominant one, and if it dies the next-weakest will be picked on, and so on down the line.

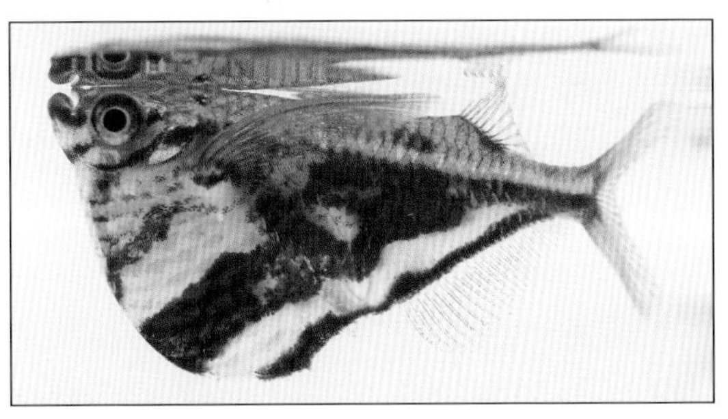

Tip 344 *Surface-dwelling hatchetfish* (Gasteropelecus *spp.*) *are liable to jump out*

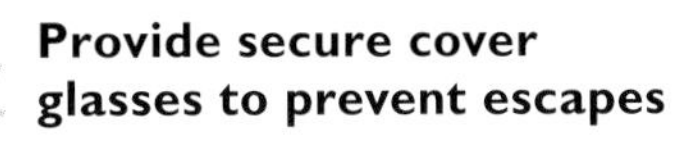

344 Provide secure cover glasses to prevent escapes

Many tropical fish are expert escape artists and can find their way to freedom through the smallest gap in the aquarium hood. To combat this, you'll need a tight-fitting cover glass which, in some instances, also needs to be heavy to stop the fish from dislodging it. Other species (for example, hatchetfish) are renowned jumpers and should never be kept in open-top tanks.

Tip 343 *A group of five-month-old blue diamond discus* (Symphysodon aequifasciatus).

345 Keep oscars amused to save your tank equipment

Large "pet" fish, such as oscars *(Astronotus ocellatus)*, can become bored with their own company if kept singly. To keep them occupied provide them with toys. A ping-pong ball may stop them from turning their attentions to more vulnerable items, such as heaterstats and internal power filters, which they can dislodge.

346 Is it true that corydoras' barbels can wear down?

It is a myth that gravel substrates directly cause the barbels of corydoras catfish (*Corydoras* spp.) to wear down. What happens is that detritus gets trapped between the grains and attracts harmful bacteria, which in turn infect the barbels. A river sand substrate is better for corydoras, as they can root around in it as they would in the wild.

347 Help your New Guinea rainbowfish to feel secure

The New Guinea rainbowfish *(Melanotaenia praecox)* will show its best colors only if it is kept in mixed-sex groups of six or more in a well-planted tank, which imparts a feeling of security. In such setups, the brighter males will constantly be displaying to the females, but any aggression will be diluted through numbers.

Tip 346 *Gravel substrates may harbor harmful bacteria.*

Tip 347 *Rainbowfish display shimmering colors.*

The flying fox (Epalzeorhynchos kalopterus)

The Siamese algae-eater (Crossocheilus siamensis)

348 Betta are splendid but shortlived

Don't be upset if your betta *(Betta splendens)*, also called Siamese fighting fish, last only a few months. They are probably dying of old age, rather than from anything you are doing wrong. Adult bettas, shipped over from Southeast Asia, are already at least six months old and half their natural lifespan is behind them. A two-year-old betta is an ancient specimen.

349 Why does my clown loach sometimes lie on its side?

Clown loaches *(Chromobotia macracantha)* often lie on their sides, behavior associated with ill-health in virtually all other fish. Nobody is sure why they do this, but in the wild they like to wedge themselves into crevices for security. If their tank lacks suitable rocks or pieces of bogwood, the clown loaches try to maximize body contact with any available surface—hence the strange postures.

350 Watch out for bullying among silver sharks

In small schools of silver sharks *(Balantiocheilus melanopterus)* it is common to find that one or more stay stunted, developing overlarge heads and emaciated bodies. This results from bullying by a dominant fish: the victims expend energy during the chase but are pushed aside at feeding time. Dense planting and providing more refuges in the aquarium, plus adding more fish to the school, may improve the situation.

351 Be sure to buy the right fish for the right job

If you are after a fish to perform a specific task in your aquarium, make sure you are sold the right one. The Siamese algae-eater *(Crossocheilus siamensis)* does very well what its common name suggests, but because it looks so similar to the flying fox *(Epalzeorhynchos kalopterus)*, even dealers confuse the two. This has resulted in the flying fox gaining an undeserved reputation as a scourge of algae.

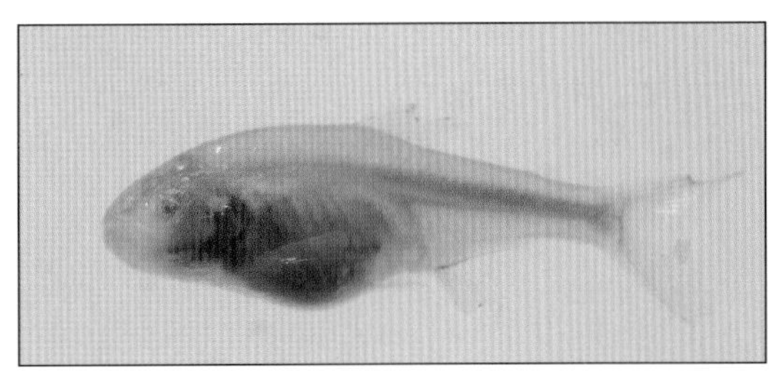

Tip 352 Astyanax fasciatus mexicanus, *the blind cave tetra.*

352 Blind fish can still see their way round the aquarium

The blind cave tetra *(Astyanax fasciatus mexicanus)* has no special sensory powers, yet fends very well for itself in the company of peaceful tankmates. Likewise, if any non-predatory, normally sighted fish loses one or even both eyes through disease or accident, it will probably still be able to lead a near-normal life. The lateral line will "see" to that.

353 American flagfish earns its place in a community tank

Don't neglect the old favorites, they can be every bit as fascinating as the latest fashionable imports. Take the American flagfish *(Jordanella floridae)*. It's a killifish that's shaped like a platy *(Xiphophorus maculatis)* but behaves like a cichlid, with the male brooding eggs and larvae in a pit. The sexes are easy to tell apart, and it makes an ideal and inexpensive community fish. Ask your retailer to order some for you.

Tip 353 *The pattern of stars and stripes on the American flagfish* (J. floridae) *account for its common name. These are males.*

354 Glass catfish are transparent beauties

Not all catfish are bottom-dwelling and nocturnal. So-called glass cats *(Kryptopterus bicirrhis)* are schooling midwater swimmers, and the two species that best merit the name, although superficially similar, belong to different families. The Nile glass cat *(Parailia pellucida)* has four pairs of barbels and comes from Africa, while the Southeast Asian glass cat (*Kryptopterus* spp.) has only one pair of barbels. Both are fine in communities of small, peaceful tankmates.

355 Are my gouramis really kissing each other?

Attribute human characteristics to fish and you will invariably misinterpret their behavior. The kissing gourami *(Helostoma temminckii)* is popular for its seemingly affectionate mouth-to-mouth displays with others of its kind, but in reality these are relatively harmless trials of strength between rival males.

356 From beautiful baby to predatory adult

Beginners are often lured into purchasing inappropriate fish for community tanks because the juveniles bear no resemblance to the adults. The classic example is the red snakehead *(Channa micropeltes)*, beautifully striped and seemingly inoffensive when imported as a 2½-inch (6 cm) baby, but it grows into a huge, drab and savage predator. So if in doubt—don't buy!

Tip 355 *It is not possible to tell the sexes apart in kissing gouramis.*

357 Not all fish adopt the same swimming posture

Learn to recognize what is the normal swimming posture for the fish species you keep. It can save you needless worry. Headstanders (*Anostomus* spp.) swim head-down, while the penguinfish *(Thayeria obliqua)* adopts a head-up posture. Swimming upside-down (inverting) is perfectly normal for the upside-down catfish *(Synodontis nigriventris)*, which even has a dark underside and a pale dorsal surface—the reverse of the usual anti-predator strategy.

358 Provide plant cover for surface-swimmers

If you don't plant your tank, you will notice that fish which normally inhabit the upper layers tend to stay further down in the water, because they lack the necessary feeling of security. If you can't grow real plants successfully don't worry; plastic will still provide the required cover to bring the fish up into the world.

359 Chinese algae eaters can become a nuisance

Some tropical fish remain popular even though their behavior is shown to be suspect, or because they are wrongly believed to carry out a useful function in the aquarium. A good example is the so-called Chinese algae-eater *(Gyrinocheilus aymonieri)*, which in adulthood is far more likely to attach itself to flat-bodied tankmates, such as discus *(Symphysodon aequifasciatus)*, and rasp away at their scales than eat algae.

360 Which are the most peaceful community fish?

True community fish remain compatible with others at all stages of their lives, but it is not always easy to assess this when you are buying them. Most farm-bred species are imported as youngsters, which have yet to develop possible aggressive or territorial traits associated with sexual maturity. As a rough rule of thumb, fish that exercise no brood care (egg-scatterers) are likely to be the most peaceful. They include barbs, danios and tetras.

Tip 357
The penguin tetra (Thayeria boehlkei) swims with its head up.

The upside-down catfish (Synodontis nigriventris) lives up to its common name.

Better Compatibility

361 One of my fish is annoying the others in the tank

In a community tank, all the occupants should ideally coexist without major problems. However, compatibility is not an exact science, and some fish will display rogue behavior not typical of their species; for example, Chinese algae-eaters *(Gyrinocheilus aymonieri)* that develop a taste for the mucus of their tankmates, or so-called "sharks" that harass anything venturing near their hideaway. The only permanent remedy is to move the offender to another tank or take it back to the store.

362 How do I cope with disruptive tiger barbs?

Some fish are badly behaved if kept singly or in ones and twos, but turn into model citizens in groups of six or more. This is true of tiger barbs *(Puntius tetrazona)*, which, having developed the school mentality, will not nip the fins of tankmates. Instead, they become preoccupied in sparring with or displaying to their own kind.

363 Established tankmates will be safe from predation

Young fish that grow up together tend to regard one another as "family." The classic scenario is angelfish that grow up with small tetras. The angel could easily eat the tetras, but does not. It's a different story, though, if the tetras are added to a tank containing an adult angel, which will thank its owner for providing it with a treat of fresh live food!

364 Night prowlers may be to blame for fish losses

Be suspicious of all catfish (except corydoras and the various plecs). When smaller fish start disappearing, the crime is usually committed at night, when the catfish are active and their prey is asleep or off its guard. Assume that if other fish can fit into the often large mouth of a catfish, then sooner or later they will—permanently.

These brightly marked fish are best kept in small schools to protect tankmates from their fin-nipping tendencies.

Tip 362 *Tiger barbs* (Puntius tetrazona) *can be disruptive.*

365 Red-tailed black sharks can be aggressive

Some community fish are best kept singly, as they will fight with their own kind. Red-tailed black sharks *(Epalzeorhynchus bicolor)* are a good example: don't even mix them with others of their genus, as similarities in color and body shape evoke an aggressive response. Occasionally, color alone is the trigger, so that the shark will pick on any dark tankmates, including fish not remotely related to it.

Tip 366 *Male betta* (Betta splendens) *in aggressive posture.*

366 Only keep one male betta per tank

The betta *(Betta splendens)* Siamese fighting fish is a perfect community tank occupant as long as you restrict yourself to only one male; two will battle to the death. In fact, with its long finnage, this fish is more likely to be a victim than an aggressor, and it should not be kept with snappy barbs or tetras.

Tip 365 *Choose the red-tailed black shark's tankmates with care.*

367 Not all gouramis behave in the same way

Avoid generalizations: all gouramis, for example, are not the same in their behavior. Dwarf *(Colisa lalia)*, honey *(C. china)* and chocolate *(Sphaerichthyos osphromenoides)* gouramis can be extremely shy and retiring, whereas the three-spot gourami *(Trichogaster trichopterus)* can be boisterous or even downright aggressive, especially the male when in breeding mode.

368 I never see my kuhli loaches, where have they gone?

Community fish may be peaceful and easy to keep, yet still not make ideal tank occupants. Kuhli loaches *(Pangio kuhli sumatranus)*, for example, are famously secretive and can disappear for months on end in the gravel, or even hole up inside a filter. In other words, they are not compatible with you, because you rarely, if ever, see them.

Tip 368 *The kuhli loach* (Pangio kuhli sumatranus) *may prove an elusive subject in the aquarium.*

Better Compatibility

Tip 369 *Keep clown loaches* (Botia macracantha) *in a small group of three or more.*

369 Clown loaches need their own kind around them

Some fish need the company of their own kind or they never seem to thrive. It is not humane to purchase single clown loach *(Botia macracantha)* or corydoras, for just this reason. They may not be true school fish, but they are certainly gregarious and will pine without others of their species—even though they may never spawn in the aquarium.

370 Are there any cichlids suited to a community tank?

Substrate-spawning cichlids do not make good community fish. Once they decide to breed, the circular territory they claim is usually larger than the tank floor; everything else is then herded up and harassed. A few small cichlids, notably kribs *(Pelvicachromis pulcher)*, are suited to community tanks, 36 inches (90 cm) minimum, provided their spawning cave is placed at one end and not in the middle of the aquarium. Cockatoo dwarf cichlids *(Apistogramma cacatuoides)* will also thrive in a sufficiently large, planted community aquarium and will use the plants to establish territories.

371 Keep the right sex ratios of live-bearing fishes

Compatibility can come down to sensible sex ratios. Livebearer males are constantly seeking to mate, so it is best to have at least three females for every suitor, enabling them to get a little peace by spreading the amount of attention paid to them. Otherwise, the females may die of exhaustion.

Tip 370 *Male cockatoo dwarf cichlid* (Apistogramma cacatuoides).

372 Piranha are safe in large, planted species tanks

Piranha *(Pygocentrus nattereri)* are ideal occupants of large, planted species tanks; being carnivores, they will not disturb the plants. Start with a school of juveniles and accept some losses along the way, because young piranha will snap at their schoolmates. This antisocial behavior quietens down with maturity. Piranha are another fish that need the company of their own kind: single specimens are shy and retiring.

373 Consider the temperaments of the fish you wish to keep

Certain fish species should not be mixed because of their widely differing temperaments. For example, bala sharks *(Balantiocheilus melanopterus)* are hyperactive and always on the move. They would distress more sedate tankmates, such as discus or angelfish.

374 Not all tropical fish need the same water temperature

A community tank is always a compromise, but water parameters, especially temperature, should not be way outside the normal range of any of the occupants. For example, White Cloud Mountain minnows *(Tanichthys albonubes)* prefer their water at 64°F to 72°F (18–22°C) and would not be happy mixed with bleeding heart tetras *(Hyphessobrycon erythrostigma)*, which require 73°F to 82°F (23–28°C). Ensure that your chosen fish fall within a similar temperature band.

375 Provide some shade in the aquarium for albino fish

Albino varieties of corydoras, barbs, sharks and other community fish do not appreciate over-bright tank lighting, as their eyes are unpigmented and their vision poor. Provide them with shade in the form of plants and decor, and consider spotlighting their home so some areas of the tank remain in shadow. To ensure they get their fair share of food, feed them when the room lights are on, but before you switch on any tank lighting. Reverse the process at lights out.

Tip 375 *Protect albino corydoras catfish from bright light.*

Better Breeding

376 Fish that look alike may not breed the same way

It is wise never to generalize about breeding fish, as fish that look similar may breed quite differently. For example, otos (*Otocinclus* spp.) lay adhesive eggs and swim away, but knifefish (*Hypoptopomas* spp.) spawn in caves and the male guards the eggs. Certain species do not always breed in the way the books say. Keep an open mind and be prepared to experiment with your setup. Crucially, where possible, check that both sexes are present.

377 Read about breeding conditions before you start

Read about the natural habitats of the fish you wish to breed. Many fish will live quite happily in a community tank. However, when it comes to breeding, you may have to help them by providing conditions that are somewhat more specialized, perhaps in a separate breeding tank. For example, breeding fish often need water parameters more similar to those found in their natural habitat.

378 How do I select good breeding stock?

This is a very important aspect of breeding fish. Only use fish with strong, robust bodies and perfectly formed fins that are carried well. Look closely at the finnage, as it is very easy to overlook a missing pelvic fin. Avoid fish with deformities such as a crooked spine. This could be a genetic fault that will be passed on to the young.

Male

Tip 379 *Even young bleeding heart tetras* (Xiphophorus maculatus) *show clear sex differences.*

Female

379 How do I sex my egg-laying fish for breeding?

It is easy to sex many of the adult egg-laying species, such as barbs, tetras, rasboras and danios, because males are more colorful and slimmer than females. Males also often have extended finnage. When they are young, it is best to buy a group of at least six juvenile fish, as sex differences may not be apparent at this stage. The larger the group the higher the chances of having both sexes. This also applies to species that are difficult to sex even when they are of breedable age.

Tip 378 *A healthy clown loach*

380 Sexing live-bearing poeciliids is straightforward

Live-bearing poeciliids are easy to sex, as the male's anal fin is modified into a rodlike structure called a gonopodium. Male swordtails, develop a swordlike extension to the base of the caudal fin, while females are usually larger, with a normal anal fin. Wild males are often more colorful than females, but cultivated female swordtails, platies and mollies can be as highly colored as males.

Male platy with gonopodium.

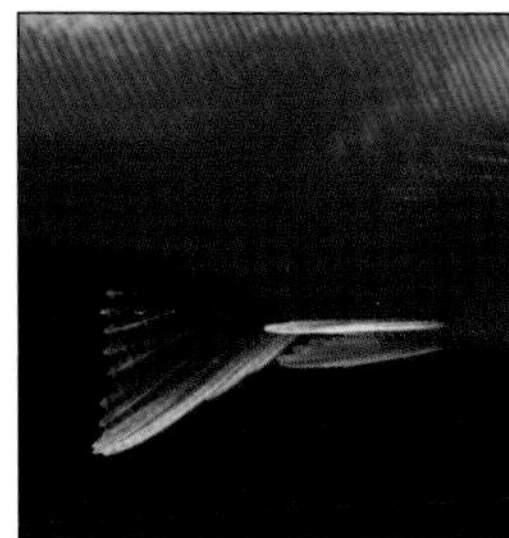

Female has a normal anal fin.

381 Separate the sexes before spawning egg-scatterers

Several weeks before attempting a controlled spawning, separate the sexes of egg-scatterers, such as many barbs, in order to condition them. Choose the most brightly colored male and the plumpest female; females in the community may be driven to extrude their eggs and are therefore slimmer. If separated from female company, the male will be excited and ready to spawn when reintroduced to the female.

382 How can I encourage my fish to spawn?

Many fish, including barbs, danios, minnows and rasboras, will be triggered to spawn when early morning sun strikes the tank. Position the breeding tank where it will catch the first light. Species such as corydoras can often be induced to spawn by carrying out a cool water change.

Breeding methods of popular fish

Fish use a wide range of breeding strategies, some of which are highly specialized. The most common methods seen in aquarium fish are listed below. Cichlids practice parental care and have specific needs in the aquarium. For advice on breeding cichlids, see Tips 418 to 427.

Egg-scatterers and egg-depositors

Many of the tetras, barbs and minnows scatter their eggs all over the place. Others, such as corydoras and many rasboras, are egg-depositors that place their adhesive eggs on plants, on the sides of the aquarium and on other smooth surfaces.

Bubblenesters

Bubble-nesting male fish construct a nest of bubbles prior to mating. Having mated, the fish place their fertilized eggs in the nest, where they are guarded by the male until the fry are free-swimming. Many of the commonly kept gouramis and other labyrinth fish use this breeding method, such as honey *(Colisa chuna)*, pearl *(Tichogaster leeri)*, kissing *(Helostoma temminchi)* and dwarf gouramis *(C. lalia)* and betta *(Betta splendens)*. Some bettas do not construct bubblenests but hold the eggs in their mouth until the young are ready to swim out. These are less common in the trade, but include *B. brederi* and *B. dimidiata.*

Livebearers

In female livebearers, following internal fertilization and a gestation period, well-developed live young (as opposed to eggs) are produced. Livebearers include platies, mollies, swordtails and guppies.

Better Breeding

383 Choosing a breeding tank and furnishing it

Tank size is related to the size of the fish and how they use the space when spawning. The usual tank sizes used for breeding are 24 x 12 x 12 inches (60 x 30 x 30 cm) for smaller species and 36 x 12 x 12 inches (90 x 30 x 30 cm) for larger ones. Plants are best placed in pots, as some tanks are bare-bottomed and others are temporary setups.

Tip 383 *Keep plants in pots in breeding tanks.*

384 When do fish become mature enough to breed?

Individual fish mature at different rates. As a guide, live-bearing platies and swordtails need fully developed gonopodiums to breed, which usually develop at two to four months. Some males are late-developers; they are usually the large ones that you need to use as breeders to obtain the best fry. Female livebearers, on average, are usually breedable at three months old. When mature, male egg-layers such as barbs develop more intense coloration and extensions of the finnage. Females fill with eggs and their rounded bellies indicate they are ready to spawn. Corydoras and many small egg-layers reach adult color in three months. Larger fish species can take longer to reach breeding age. However, some rainbowfish such as, *Melanotaenia lacustris,* become mature enough to breed within their first year, when they are 2 inches (5 cm) long, well before they reach adult size of 4$^{3}/_{4}$ inches (12 cm).

385 Altering water parameters for breeding purposes

In order to breed some species that otherwise live happily in a community aquarium, you will need to alter the water chemistry very gradually. For the best results, place the parent fish in their accustomed tank water conditions in a separate tank and change the pH and hardness levels very slowly until they reach the required readings.

386 What supplementary foods should I offer breeding fish?

It is best to add live foods to the diet of potential breeders; it is an important part of conditioning the fish. Brine shrimp, whiteworms and grindalworms can all be cultured and are safe foods. Feed these in addition to flake and frozen foods. Herbivores require extra vegetable matter. Feed the fish more frequently, but make sure that all the food is eaten quickly and remove any waste to avoid pollution.

Creating a peaty aquarium

A *Crumble a 2-inch (5 cm) thick layer of peat onto the water surface. Use either rainwater filtered through carbon or R.O. (reverse osmosis) water mixed with tap water to achieve the correct hardness.*

387 What filter should I use in a breeding tank?

Power filters with strong currents can disturb bubblenests and suck up small fry. Undergravel filters can pull babies down when fry are resting on the bottom, but they are not widely used in breeding tanks. Box filters can trap small fry in the wool. Air-powered sponge filters are ideal, as they provide gentle filtration and the biological surface of the sponge becomes a feeding ground for fry. (See Tip 287.)

388 Spawn rainforest fish in soft acidic water

Cardinals *(Paracheiron axelrodi)* and neon tetras *(P. innesi)*, which live in Amazon rainforest rivers and streams, need special conditions to spawn. Provide softer water with a lower pH than that found in a community tank. A peaty substrate will also help. Alkaline water can render inviable the sperm of fish originating from soft-water habitats, so eggs are never fertilized. Provide subdued lighting or the eggs will not hatch.

389 I've heard that earthworms are a valuable food

The highly nutritious value of earthworms makes them excellent food to condition larger fish for breeding. On a wet night, shine a flashlight over the soil and lawn in your yard. You'll need to act fast to catch the large worms lying on the surface, as they are partly buried and quickly retreat underground. Store them in a peat-filled bait box. Worms can also be bought from fishing tackle stores.

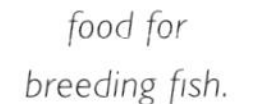

Tip 389
Earthworms, here as sold for fishing bait, are a nutritious food for breeding fish.

B *The peat will initially float on the surface and may take a week or so to sink to the bottom. Stirring it every day and squeezing pieces that are full of air helps to make it sink more quickly.*

C *The acidified water contains many beneficial trace elements. You can siphon it into another container for use with fish that require soft, acidic water or use the tank for substrate spawners or characins that are sensitive to light.*

Better Breeding

390 I've conditioned my fish but they still won't spawn. Why?

Sometimes a large female does not spawn, even after good conditioning. This usually means she is too old. Young mature adults make the best breeders. To ensure that your fish are the right age, it is best to buy young fish as a group and grow them up. Then you will know exactly how old they are.

391 Should I use a breeding trap when spawning my fish?

The idea of a breeding trap is to allow eggs or newborn fry to drop through the openings and be safe from predation by their parents or other fish. Many commercially available traps are quite small, but you can easily make your own using plastic-coated wire mesh. The mesh should be large enough to allow eggs or babies through, but not the adult fish. Form it into a basket that is a little smaller than the breeding tank and suspend it in the tank by two pieces of wire. The top of the basket should protrude 1 inch (2.5 cm) or more above the water surface to stop fish from swimming out or into the trap over the edge. The trap and the breeding fish can be removed after spawning is complete, leaving the eggs or babies to develop undisturbed.

Tip 391 *A breeding trap for egg-layers.*

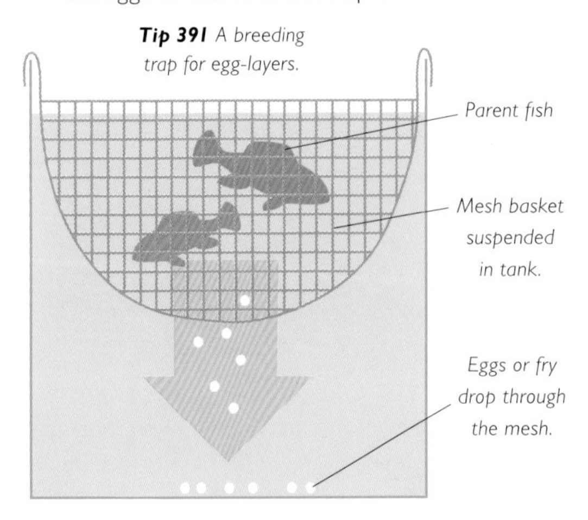

Tip 392 *A breeding trap for livebearers.*

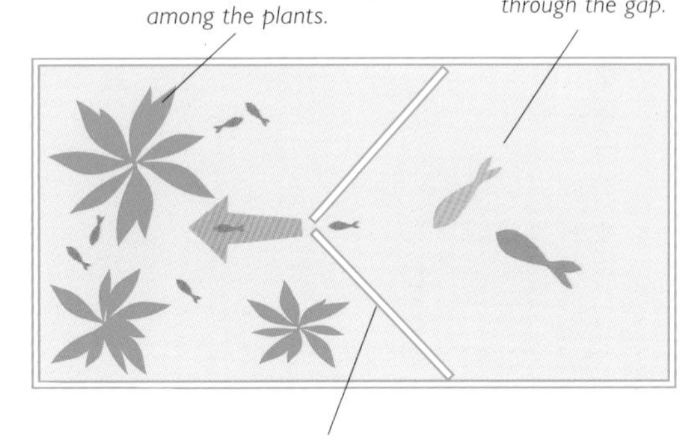

392 Making a breeding trap for livebearers

To make a V-trap for live-bearing fish, silicone two pieces of glass vertically to opposite sides of the breeding tank, leaving a small gap at the apex of the V, wide enough for the fry to swim through but not the parents. Add some plants on the side that will receive the fry and place the parents in the empty area. Livebearers that swim to the surface at birth circle around the edges looking for cover, find the gap and swim through to the plants on the other side.

393 Maternity tanks for livebearers

Female livebearers usually give birth every four weeks. About a week before she delivers, move a gravid (pregnant) female livebearer into a 12 x 8 inch (30 x 20 cm) tank filled with Java moss *(Vescularia dubayana)* and water from her home tank. Keep an eye out for the young and remove the female after delivery. If the brood is large, move the fish into more spacious accommodation as soon as possible; small broods can be left where they are for a day or two.

394 A floating plant can substitute for bubble-nesting fish

The armored catfish *(Megalechis thoracata)* and port hoplo *(Hoplosternum thorocatum)* use floating plants to hold their bubblenests together, but a simple substitute in a bare breeding tank is a Styrofoam ceiling tile. This trick also works for some gourami species, which blow bubbles beneath the tile until it is pushed clear of the surface.

395 What is the advantage of using a spawning mop?

Spawning mops are far more convenient than real plants as a spawning medium, as you have full control of how and when they are used in a breeding setup. They can be any size or density and can be placed exactly where the fish want to spawn. Since spawning mops are not generally available, most fishkeepers make their own.

396 Using spawning mops in the breeding tank

Before use, wash out the mop in warm running water. Floating mops can be attached to a piece of Styrofoam or cork. Some species are not too fussy about the positioning of the mop, but others will not spawn if it is not in the right place. Experiment with positioning until you get it right. Clean the mops after use in a strong salt solution and then rinse them thoroughly in warm water.

Making a spawning mop

A *Wind green nylon yarn around a piece of cardboard or a book until you have about 30 strands. Cut off the surplus.*

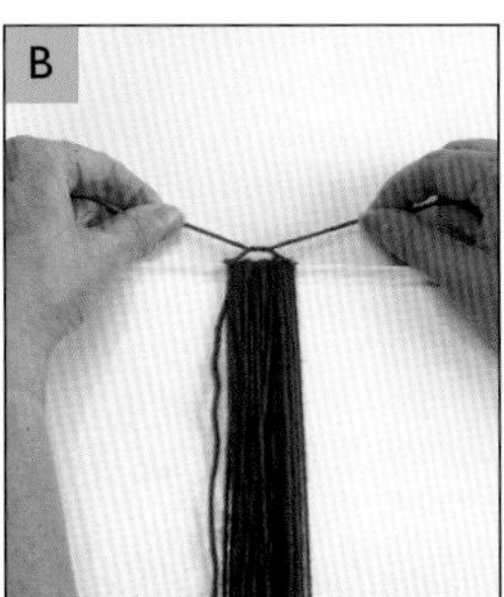

B *Cut another piece of yarn about 8 inches (20 cm) long from the ball and thread it under the strands. Secure the strands with a tight knot.*

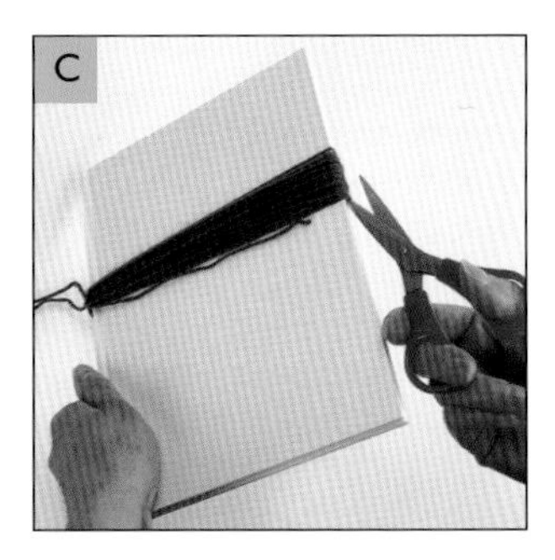

C *Turn over the cardboard or book and cut the strands at a point opposite your knot. You now have a spawning mop.*

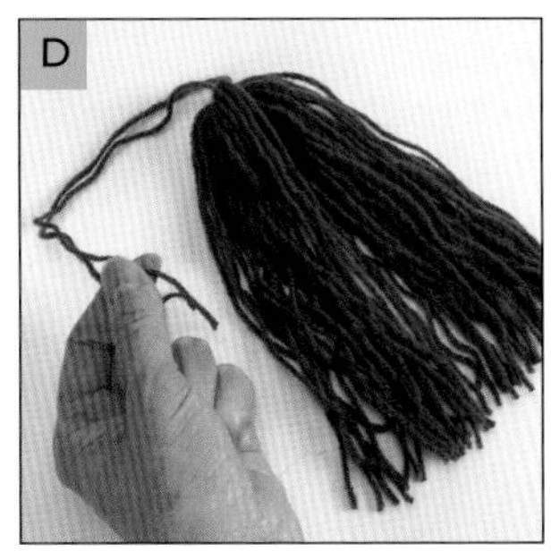

D *The long ends of nylon securing the mop strands can be used to tie the mop to a cork or to suspend it from the surface of the aquarium.*

Better Breeding

397 What are the best plants for breeding tanks?

Cabomba is excellent for egg-layers, while riccia *(Salvinia auriculata)* is ideal for surface egg-layers and livebearers. Java moss *(Vescularia dubayana)* is a favorite medium for spawning and extensively used by livebearer enthusiasts. Broad-leaved plants, such as cryptocoryne, are favored for breeding egg-depositors, such as harlequin rasboras *(Rasbora heteromorpha)*. Java fern (*Microsorium* spp.) is also popular for the same purpose.

Tip 397 *Egg-laying fish appreciate the finely branched leaves of* Cabomba caroliniana.

398 Do I need to collect the eggs as they are laid?

Egg collection is more suitable for spawners that lay low numbers of large eggs over several days, such as small rainbowfish. (Most spawnings take place in the early morning.) Eggs laid in a tight bunch could fungus, so pick them out gently with your fingers every day and spread them across the floor of a container of tank water. Where many small eggs are laid in a batch, remove the adults or transfer mops to a rearing tank. Gently aerate the mop area.

399 Prepare growing-on tanks for developing babies

Start the fry off in a small aquarium where they can easily reach their food. For the developing young, however, the more water they come into contact with, the faster they will grow. This means placing them in a larger tank. .When the babies measure half an inch (1.25 cm), start to make 10 percent water changes every few days using similar water parameters. Water changes are necessary to keep the tank clean and sweet, as fry need four feeds or more a day and there could be a lot of hungry fry. Pollution is the major issue as the fry grow and eat more. After a few weeks, increase the water changes to 25 percent. Use a small air-powered filter in the tank.

400 How can I culture infusoria for fish fry?

Infusoria, tiny organisms that live in water and feed on rotting plant matter, are the ideal first food for tiny fry and you can cultivate your own supply. Simply fill a glass jar with aquarium water, drop in a small piece of lightly boiled potato and place the jar on a window ledge. In a week the water will become cloudy with infusoria and will be ready to add to the aquarium. Start a new culture daily, as you will need enough for a week or so.

Tip 400 *Culturing infusoria for fish fry.*

401 Feeding fry on infusoria and brine shrimp

Infusoria are an essential first food for a wide range of aquarium fish, otherwise they will starve. Prepare cultures in advance as they take a week to mature. Liquid fry foods are also obtainable and are easy to use. Follow the manufacturer's instructions, being very careful not to overfeed. After infusoria come newly hatched brine shrimp. Many species can take these as a first food, but it is a good idea to offer a little infusoria to start with, as well as brine shrimp. That way, if the fry are not big enough to handle the larger food, they will not starve to death. Dried fry foods are also available, or you can used crushed, good-quality adult flake. Either way, fry need foods with a high-protein content. Feed them little and often—four times a day as a general rule.

Tip 403 *Prepare a microworm culture for baby fish.*

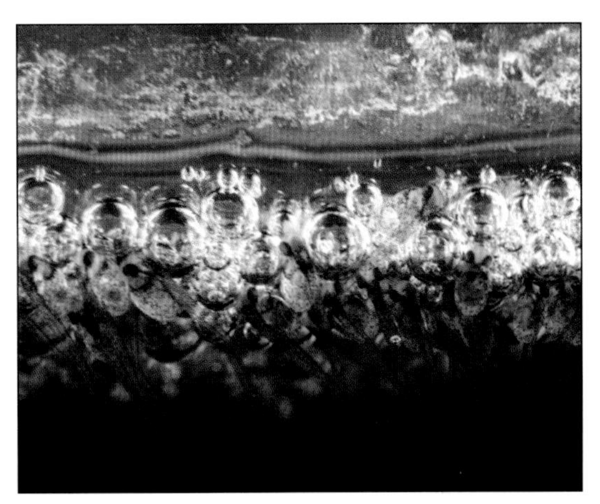

Tip 401 *Tiny paradisefish* (Macropodus opercularis) *fry need infusoria as a first food.*

402 How do I know if I am feeding enough infusoria?

Take great care when feeding infusoria. Feed too little and the fry will starve to death. Feed too much and the water will become polluted, killing the fry. Achieving the right balance can be tricky. As a guide, pour a small amount of the infusoria-filled water into the aquarium (the amount really depends on the number of fry you have to feed) and when the water clears add more. Commercially produced liquid fry foods are available. Exercise the same caution when adding these to the aquarium.

403 How do I culture microworms for baby fish?

Microworms make a good food for baby fish, and you can obtain a starter culture by mail order. Take a well-scrubbed 9-ounce (250 g) food tub and punch a few pinholes in the lid. Cook a little unflavored instant oatmeal using water only, allow it to cool and spread a half-inch (1.25 cm) layer over the base of the container. Drop a teaspoonful of culture in the middle and store the culture at 75°F (24°C). After a week, worms will crawl up the sides of the tub. You can wipe them off and feed them to baby fish.

404 Culturing brine shrimp as a fry food

Dried brine shrimp eggs are available from many aquatic outlets and are easy to hatch as a second fry food (see below). The eggs hatch in 24–30 hours, depending on temperature. You will need to keep more than one culture going to give daily feeds.

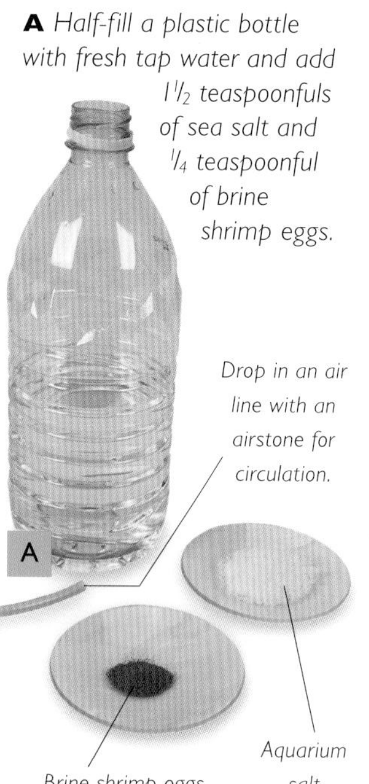

A *Half-fill a plastic bottle with fresh tap water and add $1\frac{1}{2}$ teaspoonfuls of sea salt and $\frac{1}{4}$ teaspoonful of brine shrimp eggs.*

Drop in an air line with an airstone for circulation.

Aquarium salt

Brine shrimp eggs

B *Keep the eggs in constant motion with some air. Rigid air line is best for this, but flexible tubing will also work if you position it properly. Remove the air line after 36 hours and leave the water to stand for 30 minutes.*

C *Siphon off the shrimps collected in the bottom into a jug. Separate the shrimps by pouring them through a funnel lined with coffee filter paper. Rinse them with fresh water.*

405 Feeding the fry of herbivorous species

The fry of some herbivorous species, such as bristlenose catfish *(Ancistrus dolichopterus)*, need a good supply of green food from birth. They could starve to death if food is not available. There are species, such as dwarf sucking catfish *(Otocinclus affinis)*, that thrive on green algae, and this microscopic food is often the first food the tiny fry will take. To raise a bloom of algae, place a container filled with tank water in a window. Algae will quickly develop. Add some of this green water to the tank as needed. In permanent breeding setups you can use bright light to encourage algae growth on the tank glass and on internal decor. Later on, you can introduce crushed green vegetables, such as zucchini, cucumber, shelled peas, lettuce and spinach.

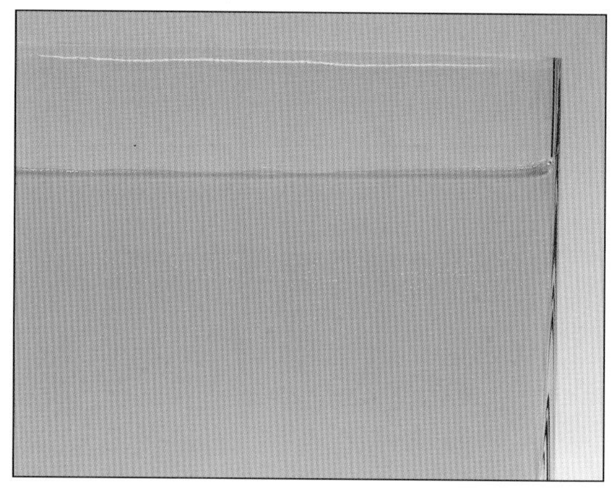

Tip 405 *Green algae as fry food.*

406 Don't bring pond foods into the breeding tank

Do not introduce pond foods such as *Daphnia* into the breeding tank, as you may introduce pests and diseases with them. Hydra that can engulf and kill fry with their long, stinging tentacles, and sluglike planaria worms that slither around at night in search of eggs and fry can be introduced on plants—don't let them in.

407 How long should I feed my young fish on fry food?

Observing the fry every day, especially when feeding them, is very important. Stages in feeding are related to growth rate, so move on to the next stage of feeding as they grow. Do not feed egg-layer larvae until they become free-swimming fry. Green water stage for some gouramis, rainbowfish and other tiny fry can last for two weeks or more. The infusoria stage usually lasts one to four weeks. Many species are large enough to take newly hatched brine shrimp as a first food. Powdered fry foods can also be given at this stage. Brine shrimp can be continued with all growing fish and adults up to 3 inches (7.5 cm). If you are uncertain when to introduce new foods, feed a little of the next stage food with the smaller food and see if it is eaten. Certain babies, such as dwarf gouramis *(Colisa lalia)*, grow at vastly different rates, some growing to twice the size of their siblings. Move these into a separate aquarium when size differences make it necessary.

Young sterba's catfish (Corydoras sterbai)

408 Be prepared to cull fish in order to control numbers

Some species produce hundreds of young in a single spawning. Where tank space is limited it is essential to cull. A general rule is to allow at least 10 square inches (62.5 cm^2) of water surface area per inch (2.5 cm) of fish, or up to 20 square inches (125 cm^2) in heavily fed fry tanks. Many losses are the result of overcrowding. Weed out any fish showing deformities, then healthy fry if necessary. It may seem cruel to feed culls to adult fish, but it's nature's way of keeping fish stocks at a level the environment can sustain.

409 Do not breed from fish with deformities

Deformities manifest themselves as fry grow, and affected fish must be culled. Crooked spines and missing fins are things to look out for. Most fish use their swim bladders to maintain equilibrium, as this organ gives the fish buoyancy. Some fry are unable to rise from the bottom or have great trouble maintaining equilibrium. Closely observe any affected fish to see if this is a temporary condition or that of a belly slider. This condition results from inbreeding, when fish are unable to rise from the bottom and only wriggle. Cull affected fish. However, some bottom-dwellers, such as many of the gobies, have little use for the swim bladder so it is much reduced or non-existent. Do not cull these fish as bottom-hugging is normal for them.

410 Tank hygiene is vital for a healthy breeding tank

Cleanliness is essential to minimize the risk of introducing pests and diseases into the breeding tank. Thoroughly clean the tank with warm water and wipe it with a clean cloth. Do not use detergents or soap. If you intend adding any plants, dip them in warm salty water, but be sure to rinse off all salt afterward. Scrub pots and other decor, and be sure to rinse any new substrate material so that it is clean and free of dust and debris.

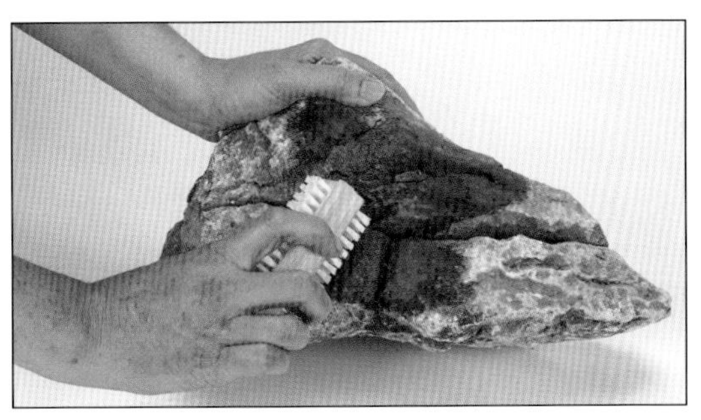

Tip 410 *Scrub rocks clean of dust and debris.*

Better Breeding

411 Keep a watchful eye for signs of velvet

Gold dust, or velvet *(Piscinoodinium)*, is a protozoan parasite that can decimate broods of fry. It appears as a dusting of tiny brown spots and can be introduced on parent fish. Use a proprietary remedy as soon as symptoms are noticed or, better still, treat your broodstock and the tank water prophylactically before spawning is attempted. Overdosing can easily kill fry; always follow the manufacturer's instructions exactly.

A double layer of glass marbles makes a good spawning substrate for danios and other egg-scatterers.

Tip 412 *Zebra danios* (Brachydanio rerio) *tend to spawn just above the substrate.*

412 What is the best spawning medium for zebra danios?

Java moss *(Vescularia dubayana)*, spawning mops, marbles covering the base of the tank or large roomy traps are all used in breeding tanks for egg-scatterers, such as barbs, danios and rasboras. An adult pair can produce up to 400 eggs. Many of these fish are rabid fry-eaters, so remove the parents when spawning is completed.

413 Do all live-bearing fish eat their fry?

Many commonly kept cultivated livebearers, such as guppies *(Poecilia reticulata)*, platies *(Xiphophorus maculatus)* and swordtails *(X. helleri)*, eat their fry, so ideally transfer gravid females to a separate breeding tank before they give birth. However, not all species of livebearer eat their fry; for example, least killifish *(Heterandria formosa)* thrive in a family group, and many other wild livebearers are the same, including blue limias *(Limia vittata)*. They are happiest when kept in well-planted tanks with their own kind.

414 Cave-spawning catfish need a permanent breeding setup

Bristlenose catfish *(Ancistrus temminckii)* need caves made of clay pots or rock structures in which to lay their eggs. These are protected by the male, who continues to supervise the fry until they venture out. To encourage the algae growth on which the fish graze, site the tank in a spot where it receives bright light for part of the day, or increase the intensity of the tank illumination for a few hours. For fish that breed in this way, it is best to keep them in a tank that has been permanently set up to suit their spawning style.

Include plants such as cabomba and vals.

Tip 414 *Cave-spawning catfish require clay pots, rocks and bogwood with holes.*

415 Egg-depositors seek out a variety of breeding sites

Caves and hollows under bogwood are favorite spawning sites for egg-depositors, such as catfish, so make sure that the setup is appropriately furnished. Many species, such as harlequin rasboras *(Rasbora heteromorpha)*, prefer to spawn on Java fern and other broad-leaved plants. Danios and minnows are not so fussy, scattering their eggs on smooth surfaces, such as the sides of the aquarium.

Tip 416 *The male thick-lipped gourami* (Trichogaster labiosus) *builds the bubblenest.*

416 Bubblenesters use floating plants at the surface

When breeding bubblenest builders, such as dwarf gouramis *(Colisa lalia)*, select a male that is already blowing a few bubbles and a female plump with eggs. Do not use any filtration in the breeding tank, as it can disturb the bubblenest. Provide some floating plants for nest construction and a tight-fitting cover. The very tiny fry will need feeding on infusoria longer than most.

417 What is the safest way to spawn bettas?

When spawning bettas *(Betta splendens)*, it is a good idea to incorporate a sliding glass divider in an aquarium measuring 24 x 12 x 12 inches (60 x 30 x 30 cm). House the ripe female in a portion of the tank (about 6 x 12 x 12 inches or 15 x 30 x 30 cm), as she may need to be introduced several times before spawning takes place. Observe the fish closely throughout the spawning activities to ensure the survival of the female during the pre-spawning mayhem. Remove her immediately after spawning is completed.

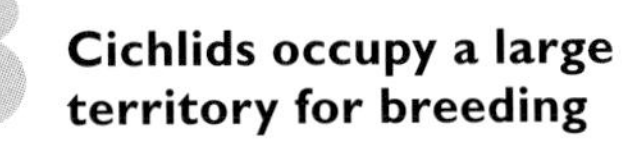

418 Cichlids occupy a large territory for breeding

Substrate-spawning cichlids occupy a roughly circular territory, centered on their spawning site and with a diameter 10 or more times the length of the male. Because tanks are usually rectangular, part of the circle is outside the glass, but the pair will occupy a tank length corresponding to normal territory diameter. Unless the tank length is substantially greater than territory diameter, the cichlids should have the tank to themselves.

Tip 418 *Cichlids occupy a circular breeding territory.*

Better Breeding

419 Tank dividers have many uses when breeding cichlids

A divider is invaluable when breeding cichlids in order to separate male from female, adults from fry or the cichlids from other fish in the event of aggression. It may be clear or opaque, depending on whether you want the fish to see one another. So when setting up your tank arrange the decor (leave gaps!) so that a divider can be used if required.

420 How should I set about pairing my cichlids?

Never assume that male cichlid plus female cichlid equals a pair. Pairing is often a prolonged process, and two unpaired, territorial cichlids in a confined space can mean a battle and the death of one. Likewise, a newly added potential mate may instead be viewed as an invader and attacked. Use a divider for several weeks while the two fish settle in and establish their separate territories.

421 How do I prevent a male cichlid from attacking a female?

When pairs of large cichlids are kept alone (usually necessary), the male may attack the female in the absence of any other target for his territoriality. A target fish can be used to avoid this, either protected by a clear divider or in an adjacent tank. Another large cichlid is ideal. The target fish must always have enough space and a hiding place.

422 Do not let breeding cichlids undermine tank decor

Many cichlids dig as a necessary part of their breeding behavior (for example, to make nursery pits for fry) and must have substrate that permits this. No substrate may mean no breeding! However, make sure rocks cannot be undermined, and that equipment is firmly secured off the bottom so it cannot be moved or buried. Plants can be grown in pots or between rocks to avoid uprooting.

Tip 421 *A male green terror cichlid* (Aequidens rivulatus) *mouthing a female.*

Tip 425 *A pair of rams guarding their eggs laid on a smooth stone.*

423 Cichlid fry are large enough to accept brine shrimp

The most common causes of cichlid fry losses are pollution of the water and starvation, often through feeding liquid foods or infusoria, which are far too small. Almost all newly free-swimming cichlid fry can manage brine shrimp larvae or microworms. Brine shrimp offer the additional advantage that their orange color shows through the translucent sides of the fry, showing whether they have eaten or not.

424 Make sure your cichlids have peace and quiet

Brood care requires a considerable investment of energy, and cichlids may eat their eggs or fry if they feel they are wasting their time in a disturbed, insecure environment. The secret of cichlid breeding is to set up the tank to suit their needs before adding the fish, then leave them in peace apart from feeding and (non-disruptive) water changes. Be patient and never fiddle!

425 Cichlid spawning sites need to be smooth

Cichlids rarely choose a rock with a rough surface for spawning. They need to clean the site with their mouths, and their lips are delicate. Furthermore, a rough surface is more difficult to clean and more likely to collect sediment around the pollution-sensitive eggs. So always provide some smooth stones for open-brooders, and include some slate or similar material when building tufa rockwork for cave-spawners.

426 Moving mouth-brooding females between tanks

When moving a mouth-brooding female to another tank, never lift her out of the water. Use a large net, raise it to the surface, and take a plastic bowl to lift the fish out with some water. Then cover the bowl (say, with the net), carry it to the new tank and float it for a minute or two to equalize the temperatures.

427 How to reintroduce mouthbrooder females

A mouth-brooding female returned to her community after brooding may be attacked as an intruder, so put her in a breeding trap for a couple of days. The other fish can see her but learn they can't get to her. Then sink the trap. The female can now start to venture out, but return to the safety of the trap if required.

Better Maintenance

428 Water changes—the best strategy

Small, frequent partial water changes are less likely to impact on water chemistry than larger ones carried out with longer intervals in between them. For example, if you change 30 percent a month, break this down into three equally spaced 10 percent changes. This is also a good way to keep on top of nitrate buildup.

Tip 428 *Water changes are a vital task.*

429 Prepare new water before adding it to the aquarium

To prepare new tap water for water changes in a community tank, fill white plastic buckets two-thirds full from the cold tap, add dechlorinator and mix it in thoroughly. Now slowly stir in hot water from a boiled kettle until the temperature equates within a degree or two to that of your aquarium. Sudden changes of temperature can shock fish and precipitate outbreaks of ich (whitespot).

430 Test the pH level of new water before adding it

Soft but alkaline water is occasionally produced by water companies to meet consumer regulations, but it is unstable. Water with minimal hardness and a pH of 7.9 when first drawn can revert to pH 6.8 after 24 hours, and would clearly not be suitable for Rift Lake cichlids. This emphasizes the need for new water to stand and be tested for pH before going into the tank. A buffer, such as sodium bicarbonate, may be needed to raise the pH to the correct level, or you can add a saltwater calcium supplement or coral gravel to a power filter.

431 Large fish need large water changes

Large, frequent water changes are necessary if you keep carnivorous fish such as piranha or the African pike characin *(Hepsetus odoe)*. The high-protein content of the food can lead to a sudden pH crash, especially in soft water areas. A bag of buffering material, such as coral gravel or cleaned oyster shell, in an external power filter may help. Failing that, weekly partial water changes of up to 30 percent may be called for. Reducing the size and frequency of meals is another option.

Tip 431 *Pike characins* (Hepsetus odoe) *need large water changes.*

432 Making water changes in breeding tanks

Essential water changes in breeding tanks can disrupt parent fish or fry, especially in the case of bubblenesters such as gouramis. To minimize stress, siphon out water slowly through air line and add new water the same way. Place the topping-up receptacle above the tank and begin the siphon action by immersing the air line in the water and bringing the free end out, stopping it with your thumb until it's in the tank. Use a clamp on the air line if the flow rate is still too fast.

433 Automatic water changers, the pros and cons

Automatic water changers are useful labor-saving devices, especially if you have several tanks to maintain. But remember that by their very nature they are adding raw tap water to the system. Never use them to change more than 10 percent of total tank volume at a time.

The tap pump can either create suction to clean the tank or allow water to pass through to fill it.

A flow adjuster controls the speed of filling and cleaning. Keep it free of debris.

The flexible hose (shortened here) takes water to and from the aquarium.

Push the tube into the gravel. Dirt and gravel swirl inside the tube, and lighter debris is drawn off by siphon action.

Tip 434 *A siphon gravel cleaner.*

Tip 434 *Organic debris is visible in the tube.*

434 Change the water and clean the gravel at the same time

Using a siphon gravel cleaner, you can make a water change and remove algae and debris from plants at the same time. Choose a model with a built-in, self-priming mechanism so you won't need to suck on the end of the hose and risk taking in a mouthful of aquarium water.

435 Don't use a gravel cleaner to clean a sandy substrate

The best way to keep a sand substrate conditioned is to run your fingers through it. If you use a vacuuming device you will inevitably suck up sand as well as dirt. The water will temporarily cloud, but an internal or external power filter should soon clear it. Next day, clean the filter media so that sand particles do not work their way into the mechanism and cause excessive wear and tear.

Better Maintenance

436 Clean beneath the undergravel plates

Attach a length of flexible tubing to the intake of a powered gravel cleaner, and insert it down the air lifts of your undergravel filter once a week. This will enable you to suck up detritus from beneath the plates, and delay the necessity for a full strip down of the filtration system.

437 Take care not to disturb potential spawning sites

During maintenance, never push siphon tubes into rock caves or under upturned flowerpots, which could be used as spawning sites by, for example, kribs or bristlenose catfish. You run the risk of destroying unseen eggs and newly hatched larvae. The parent fish will perform their own housekeeping more gently and efficiently than you ever could.

438 Wear rubber gloves when carrying out maintenance

There is a minimal risk of fish diseases transferring themselves to humans, but if you have a cut or scrape on your hand it is always best to wear rubber gloves when carrying out any maintenance inside the tank. Some fishkeepers may develop allergies to aquatic plants, in which case gloves will also help.

439 Don't wash all the filter foam at the same time

If you have an internal power filter with a foam cartridge as the only media, cut it into two pieces and rinse each half alternately in aquarium water at every maintenance session. That way you won't wash away too many beneficial bacteria.

Washing the filter foam

Tip 439 *When renewing filter foam, run a new and an old half together for a month to allow the new piece to become seeded with nitrifying bacteria.*

A *Remove the foam from the canister and squeeze it out in the aquarium water that has collected in the bucket during the water change.*

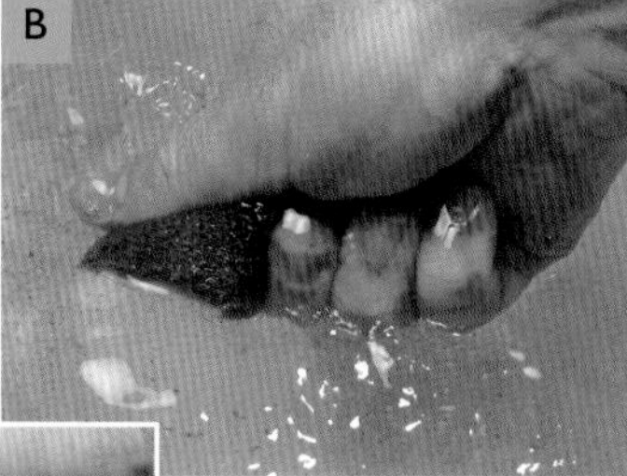

B *Keep squeezing the foam and remove any visible debris as well. It will soon begin to look a great deal cleaner.*

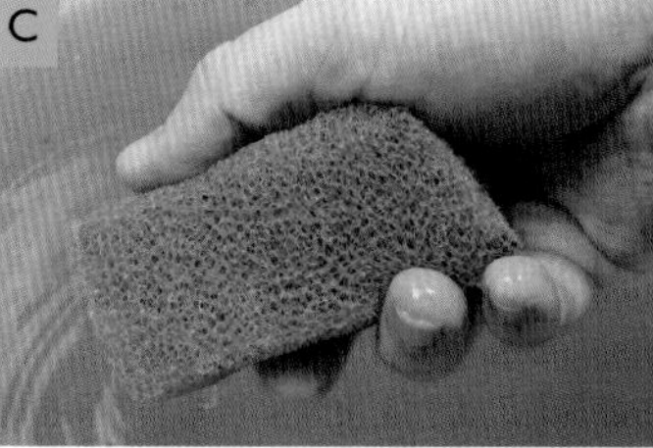

C *It is a good idea to replace the filter foam when it shows signs of not returning to its original shape when squeezed.*

440 Keep pipework clear of calcium deposits and algae

Power filter pipework can clog with calcium deposits in hard water areas, or become blocked by algae if exposed to strong light. A flexible bottlebrush with soft nylon bristles and a long wire stem (obtainable from hardware and home-brewing stores) is ideal for keeping the lines open.

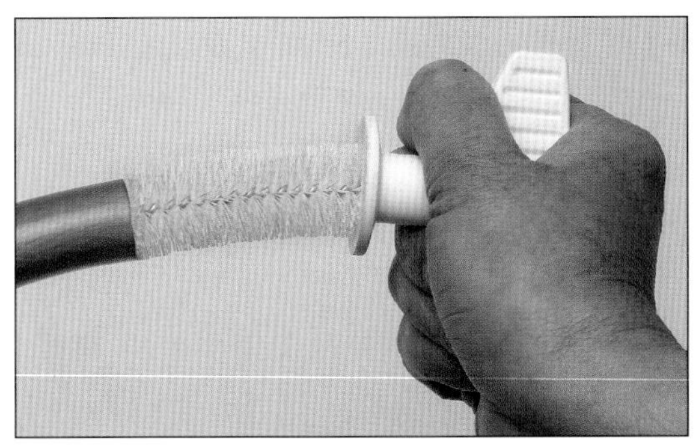

Tip 440 *Flexible bottlebrushes are ideal for cleaning pipes.*

441 Toothbrushes are a handy maintenance tool

Toothbrushes are ideal for routine maintenance tasks, such as ridding filter impellers of lime-scale, or cleaning rockwork, or getting algae off the inside corners of the tank that a magnetic cleaner cannot reach. Toothbrushes with soft nylon bristles won't scratch the glass.

442 Remove limescale from the top of the aquarium

In hard water areas, a lime-scale deposit can quickly build up around the top water level in your tank. If not removed regularly, it can be almost impossible to shift. The most efficient way do this is to hold the abrasive half of an algae magnet and use it as you would a scouring pad. Operating the magnet in the normal way will not exert sufficient pressure on the stubborn deposits to have much effect.

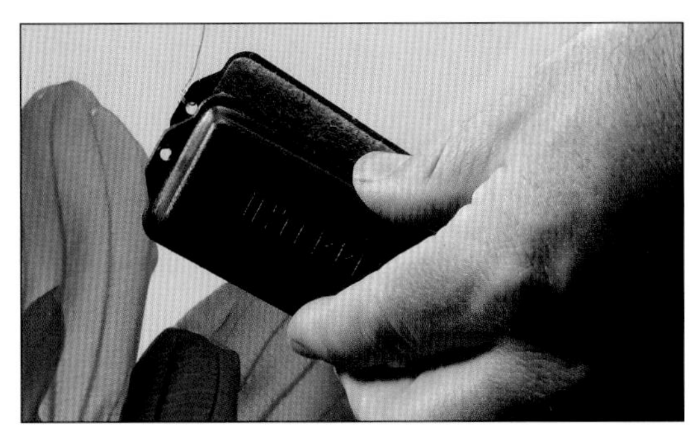

Tip 443 *With an algae magnet you can clean the inside glass.*

443 Rescuing an algae magnet if it falls into the tank

Make sure the internal half of a magnetic algae cleaner has nylon line attached to it. (If necessary, drill a hole in the casing and attach a length of heavy fishing line). That way, if the halves separate in use (as they often do), you will be able to retrieve the magnet without putting your hand inside the tank. Magnetic algae cleaners with internal air pockets will float if they become detached.

444 Leave some algae on the back and sides of the tank

Do not be too rigorous in cleaning algae from the side and back glass of the tank. Moderate growth fulfils the same function as an aquarium background, giving the fish a sense of security. The algae also houses microorganisms on which young fish in particular will feed.

445 Make sure a thermometer is too big for a fish to swallow

A tank thermometer is essential equipment and should be positioned where you can check it daily at a glance. But never use glass-bodied thermometers in tanks containing large cichlids or catfish—it has been known for them to swallow such items, causing throat damage or even gut blockages.

446 Check water circulation after maintenance

During maintenance it is easy to knock the return flows from filters and powerheads accidentally. Big fish sometimes do this too. The resulting change of water flow direction can affect surface agitation and circulation. For example, you may notice that the tank is not heating up to the correct temperature. This may be happening because a small pocket of water in a dead spot is being warmed (thus turning off the heaterstat), while the rest of the tank fails to benefit fully. So note the ideal direction of water returns and keep them that way.

447 Convenient electrical connections

The best way to connect several electrical devices for aquarium use, such as power filters, heaters, lights, etc., is to plug them into a multi-socket adapter strip (or power bar) with individual switches. Ideally, choose a surge- and spike-protected strip, as sold for use with home computers. This enables you to switch off (or unplug) each item as you carry out routine maintenance in the aquarium. Systemized aquariums in which all the electrical devices are fed by one plug look neat and unobtrusive, but if the fuse blows all the systems go down at the same time.

448 Eliminating hydra without resorting to chemicals

Hydra can be brought into the aquarium on plants. These pests resemble miniature sea anemones and can devour young fry. To eliminate them, temporarily rehouse your fish and attach wires to the terminals of a 9-volt battery, as used in domestic smoke alarms. Place the free ends of wire into the water at opposite ends of the tank and leave for 30 minutes. This is a far safer and more reliable cure than any chemical remedy.

Essential supplies

As well as making sure you have sufficient fish food, it is vital to have supplies and spare parts. Keep any small metal spares (impeller assemblies, gang valves, etc.) in a small metal tin to which you have added a few grains of uncooked rice. These will absorb any moisture and prevent corrosion before you come to use the spares.

Heater-thermostat
Suckers
Bearings, impeller and O-ring seals for filters
Filter wool or foam
Other filter media being used in the system
Clean plastic bucket or bowl
Aquarium silicone sealant
Activated carbon
Thermometer
Nets
Fluorescent bulbs
Starters for fluorescent bulbs
Battery-powered air pump
Air line and airstones
Test kits
Tap water conditioner
Filter start-up product
Remedies for common ailments, such as ich, fungus and bacterial infections

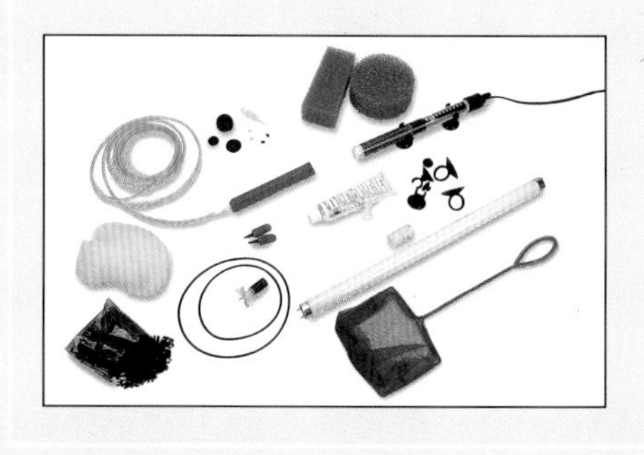

There are flatworms in my tank. Are they dangerous?

Flatworms and ostracods, tiny crustaceans (also known as seed shrimps) measuring $^1/_{125}$ to $^1/_{30}$ of an inch (0.2–7 mm) are scavengers. They are not harmful in themselves, but are a sign of inadequate maintenance and an overfed tank. If cutting down on food rations for your fish does not reduce their numbers, it may be time for a major strip down of the tank and a thorough cleaning of the substrate.

450 Maintain air pumps for reliable performance

The most common cause of failure in air pumps is ruptured or perished diaphragms, so make a point of changing these every six months, whether or not they are showing obvious signs of wear. At the same time, replace flap valves and air filters. Maintenance kits are available for most popular brands.

Routine maintenance

To keep your aquarium looking good, it is vital to carry out routine maintenance. This schedule summarizes the essential tasks you will need to carry out.

Every day

- Check the water temperature.
- Check that the internal filter and lights are working.
- Check for missing fish.
- Check for signs of ill-health or distress.
- Feed the fish, making sure that they all receive some.
- Remove any uneaten food.

Every 7–14 days

- Test the water for pH, ammonia, nitrite and nitrate levels.
- Clean the front and side glass of the aquarium.
- Remove dead or dying leaves.
- Gently disturb fine-leaved plants to remove detritus.
- Clean the substrate with a siphon gravel cleaner.
- Make a 15% water change and refill the aquarium with conditioned water at the same temperature.
- Where fitted, clean the condensation cover.

Every 4–6 weeks

- Carry out any cleaning with tank water to avoid disrupting the beneficial bacteria in the biological medium. Biological media will not need washing if mechanical prefiltering is working efficiently.
- Clean the internal filter, including the impeller and casing.
- Replace any activated carbon in the filter. Rinse it before use.
- Clean any mechanical media in external filters.
- Replace expendable filter media, e.g. filter wool.

Every 6–12 months

- Replace half the filter foam in the internal filter when it loses elasticity. Allow one month before replacing the other half.
- Replace fluorescent bulbs, even if they are still working.
- Replace the filter pump impeller.

As needed

- Replenish tablet fertilizers for aquarium plants.
- Trim tall plants to prevent them from blocking light to other plants. Use trimmings for propagation.
- Check the quarantine/hospital tank and make sure the filter is working properly in case the tank is needed for new fish or to treat existing stock with medications.

Better Healthcare

451 Can I collect pieces of aquarium decor?

Not all tank decor is safe for your fish. Avoid ornaments with sharp edges on which fish could injure themselves, or any with small holes and crevices in which they could get stuck. It can be risky to collect your own ornaments or shells; it is safest to buy items of decor from an aquatic store, as you know they will not leach any dangerous substances into your water.

Tip 451 *Make sure all aquarium decor is safe to use.*

452 Find out all you can about a fish you like before buying it

Research the health needs of species you intend to keep. Most are not fussy, but some types do have specific dietary or water requirements. There are books available listing all the common freshwater species, and many internet sites provide basic information. If you see a fish that you like in a store, check its scientific name and needs with a member of staff before buying it.

Tip 452 *Zebra danios* (Brachydanio rerio) *make good community fish.*

Tip 455 *Check your fish regularly for signs of disease.*

453 Avoid seemingly healthy fish in a suspect tank

Ensure the fish you buy are as healthy as possible. When choosing new stock, look for the same behavioral and visual signs of ill-health as you would when checking your own fish, and do your homework so you are familiar with the norms for the species. Avoid buying from a tank where several individuals seem to be unwell, even if the fish you want looks perfectly fine.

454 Prevention is always better than cure

If you follow some general rules, you can prevent most health problems. Buy healthy, good-quality stock, and quarantine it yourself when possible. Provide the right setup and environment to keep the fish happy, along with clean water of the right chemistry. Feed the correct diet for the species, and vary it where you can. Watch out for stress caused by fighting, bullying or breeding activities.

455 Recognize the normal appearance of your fish

Become familiar with the normal appearance of your fish; changes in this can be the first sign of ill-health. Look out for unusually pale or dark colors; the development of spots, ulcers or white patches; frayed or reddish fins; cloudy or bulging eyes; slimy or bloodshot skin; raised scales; and a thin, swollen or asymmetrical belly. Feeding is a good time to check for problems.

456 Recognize the normal behavior of your fish

Know what constitutes normal behavior for the species you keep, as changes in this can indicate more urgent problems. Common behavioral signs of distress are hiding, refusing to feed, resting on the surface or the bottom; breathing heavily; holding fins clamped; darting around the aquarium; and swimming with irregular movements or in odd positions. However, remember that some of these activities are normal for certain species.

457 What is the best way to handle my fish?

Fish should be moved and handled as little as possible. Netting or handling can split or injure their fins, knock off scales and damage the sticky mucus coating that helps to protect them from parasites. Leave the fish in the tank while performing regular maintenance jobs. If you ever do need to handle them, wrap them in a soft, wet cloth and be as quick as possible.

Tip 457 *You need not move fish out of the tank during maintenance.*

How does my fish work?

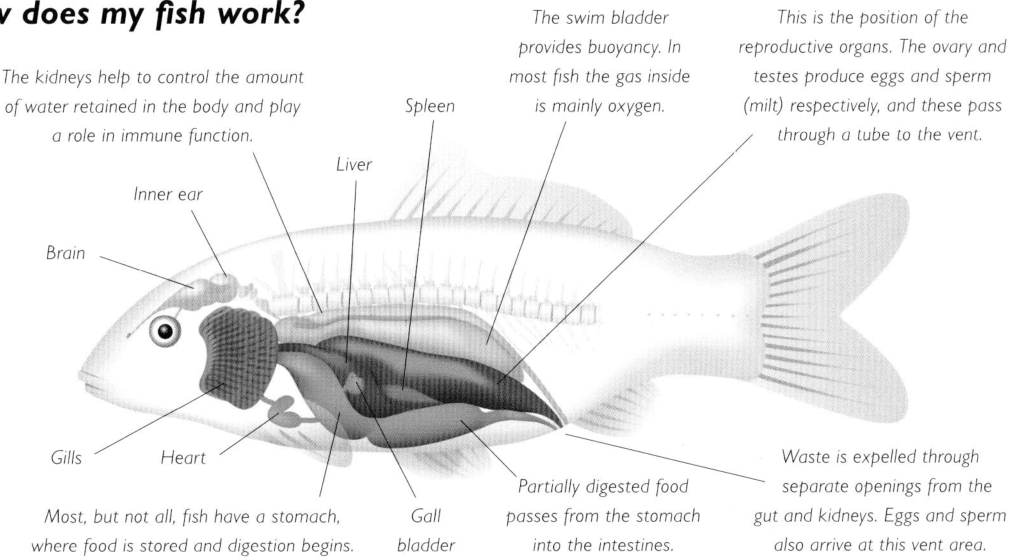

Better Healthcare

458 Look for the underlying cause of a problem

Many of the microorganisms that cause disease are permanently present on the fish's body or in its environment. Often, a fish succumbs simply because its immune system is suppressed, typically by some form of stress. Therefore, it can be a waste of time trying to work out where some diseases come from, or even reaching for the bottle of medication. Concentrate first on identifying possible underlying causes.

459 A quarantine tank has many other uses

If you can afford it, get a spare, or quarantine, aquarium. This is important for isolating new stock but has many other uses as well. It is somewhere to house the odd sick fish, bullied or harassed fish, baby fish or any tank inhabitants that cannot tolerate treatments being used in the main aquarium. It can be small, with basic equipment and stored dry when not in use.

460 Stress is bad for fish—look out for signs of disease

Because of the link between stress and disease, look out for serious health problems in the days following a serious environmental glitch. Typical scenarios are outbreaks of severe bacterial infections after ammonia or nitrite poisoning, damaged and infected fins after transport stress, and epidemics of skin parasites (especially in livebearers) after a period at too low a temperature.

461 Do I really need to quarantine new fish?

When buying new fish, even from reliable sources, it is always safest to quarantine them in your spare aquarium before introducing them to the main tank. If they do need any medication, treatment will be easier in the quarantine tank, and it can also allow them to rest and acclimatize to your water conditions. Ideally keep them isolated for four weeks, but certainly no less than one.

Temporarily house newly acquired fish in a quarantine tank.

After quarantine is completed, you can introduce the new fish into the existing community.

The main display tank.

Remove sick fish to a separate hospital tank for treatment.

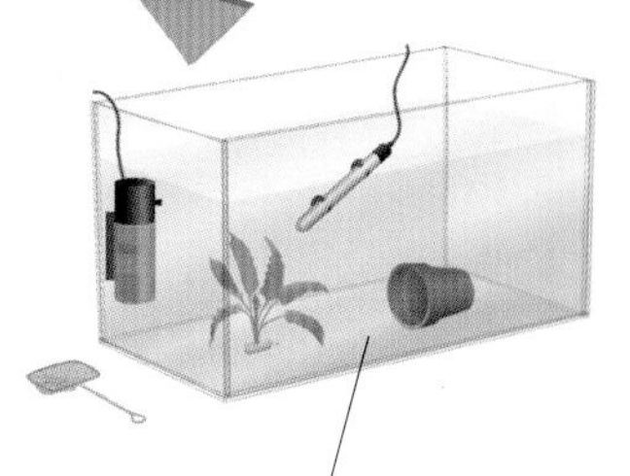

Each aquarium should have its own net and other equipment to prevent cross-contamination.

***Tip 459** A spare tank is very practical.*

A treatment tank is designed for easy treatment and cleaning.

462 One of my fish has died. Is there a problem in the tank?

It is normal to lose the occasional fish. This can be due to individual health problems or simply old age, and is nothing to worry about. However, if you lose many fish suddenly (particularly of different species), this usually indicates an environmental problem, such as water quality or temperature. If you lose many fish gradually (particularly of the same or related species), this usually indicates a disease.

463 There is usually a reason for that mystery illness

A mystery health problem may not be a disease. Check all water quality values, then do a partial water change to see if that improves things. Check that all equipment is functioning normally, and consider if anything poisonous (from medications or inappropriate decor) could have been added to the aquarium. Finally, ask yourself if the problem could be due to breeding behavior or bullying.

Air-powered, twin-sponge filter provides gentle and safe filtration.

Artificial decor provides a refuge for nervous fish.

Tip 461 *A quarantine tank setup—this one with a layer of substrate.*

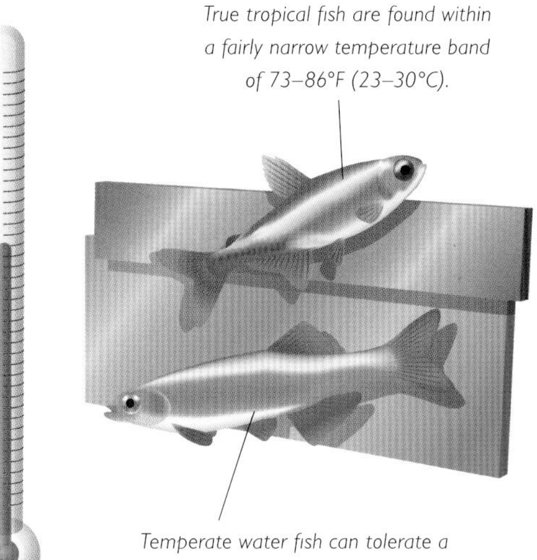

True tropical fish are found within a fairly narrow temperature band of 73–86°F (23–30°C).

Temperate water fish can tolerate a temperature range of 50–77°F (10–25°C)—a reflection of what they experience in the wild.

464 Be prepared and have contact details at hand

Occasionally, fish diseases require medications, sometimes urgently, and you can save time in an emergency if you have contact details ready. Most veterinarians don't treat fish, but some may still be able to provide advice and guidance. Call around the animal clinics in your area, before your fish fall ill, to see if any can assist. Friends, professionals and local societies are also good sources of information.

465 My sick fish is suffering. What should I do?

If a fish is terminally ill and obviously suffering, euthanasia may be the kindest option. This can be a difficult decision to make, so you may want to seek the advice of a fish society. A veterinarian may be able to administer an overdose of anaesthetic, so that the fish gradually loses consciousness and slips away without pain.

Better Healthcare

466 Ask advice about humane methods of euthanasia

If you have a terminally ill fish but do not have access to anesthetics, there are alternative methods of euthanasia. However, these must be done properly so always seek advice from someone experienced in the subject. Never flush a fish down the toilet, put it in the freezer, decapitate it or drop it into very hot or very cold water as this can cause suffering.

467 Why is my fish not swimming normally?

Occasionally, a fish can be seen repeatedly swimming with its head or tail upward, struggling to get off the bottom or to avoid rising to the surface. Sometimes this is not a disease but is caused by congenital swim bladder defects, injury, old age or simply too much dried food. If the fish is suffering euthanasia may be needed, but if it seems otherwise healthy do not interfere.

468 Consult a fish health book for more information

If you do not have access to professional advice or an experienced friend, it might be a good idea to obtain a book on fish health. However, as with reading a medical dictionary and becoming a hypochondriac, do not allow yourself to believe that your fish are suffering from all the diseases listed! The most common illnesses are ich (whitespot and other skin parasites), bacterial and fungal infections.

469 Ich (whitespot) is commonly seen in aquarium fish

Ich (also called whitespot) is still one of the most widespread fish diseases, and as it usually comes in with new stock, quarantine is still the most effective way to tackle it. Pinhead-sized white dots appear on the skin and fins, and badly affected fish may gasp due to irritation of the gills. Any species can be infected. Treat as soon as possible with an over-the-counter remedy.

Rapid gill movements could be the result of water-quality problems, as well as parasites or bacterial infection.

Slime patches are a fish's immune response to protozoan skin parasites.

Mouthrot could be a bacterial or fungal infection.

Swollen eyes can be seen in fish affected by dropsy. A tumor behind the eye could also cause the symptom.

White spots are a sign of parasite infection. Treat promptly.

470 Bacterial infections are often stress-related

Watch out for bacterial infections—they are very common, often have a rapid onset and are usually stress-related. Symptoms include fin rot, mouth rot, ulcers, bloodshot areas, pop-eye, and swelling or emaciation of the abdomen. Superficial infections can be treated with over-the-counter remedies, but more severe ones require prescription antibiotics. Try also to establish the original cause: could something have lowered your fish's resistance?

471 Healthy fish can resist fungal spores

Fungal spores are constantly present in aquatic environments. While healthy fish can resist them, they will invade damaged tissues or even intact ones if the fish's overall immune system is severely suppressed. Therefore, if a fish is stressed, has a wound or another infection, it may well also pick up fungus. Fungus appears as whitish, fluffy blobs like cotton balls and needs to be treated immediately with over-the-counter remedies.

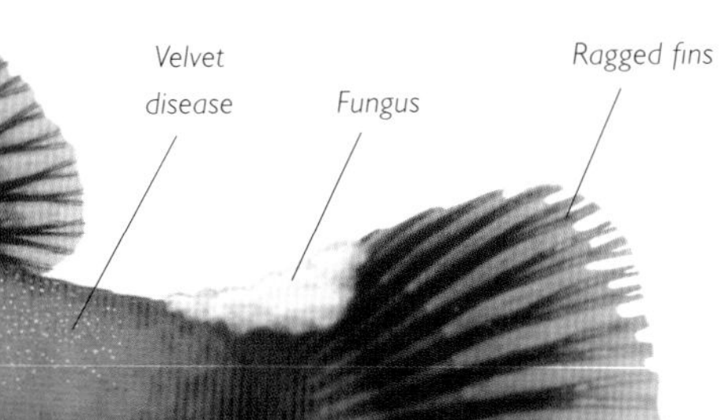

Protruding scales are symptoms of dropsy. Use an internal bacterial treatment.

472 Viral diseases can clear up without treatment

Like us, fish can occasionally suffer from viral diseases, but they do not get colds. If an individual shows one or more slow-growing whitish lumps or nodules, this is generally caused by a virus. These illnesses are not very infectious, although you may want to isolate affected animals just in case. There is no treatment, but in good conditions the problem will usually clear up by itself.

473 Skin parasites can be difficult to treat

Some species, particularly livebearers, can be attacked by minute skin parasites. Look for slimy or cloudy skin, clamped fins and rapid or heavy breathing. Sometimes the fish may have pale patches on its body. These infections can be difficult, so if an over-the-counter parasite remedy does not seem to work, contact a specialist. Stress can also play a part, so check that all your tank conditions are correct.

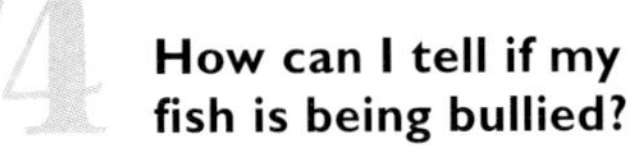

474 How can I tell if my fish is being bullied?

The symptoms of bullying can mimic those of some diseases: ragged or clamped fins, battered or slimy-looking flanks, hiding and refusing to eat. If persecution from tankmates is a possibility, try transferring the affected fish to your quarantine aquarium and simply leaving it. If it looks dramatically better with no other treatment, you probably do have a bullying problem.

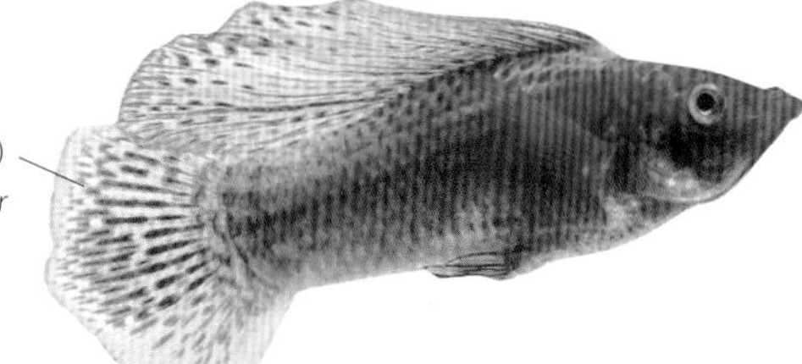

The flowing fins of a sailfin molly (Poecilia latipinna) may be a target for fin-nipping fish.

475 Mouth rot needs specialist treatment

Mouth fungus or mouth rot is an illness that particularly affects livebearers. However, do not be fooled; it is not a fungus but a bacterial infection. Symptoms consist of rough, whitish patches, especially on the mouth. It can resemble either parasitic infections (but it takes hold much more quickly) or fungal disease (but it lacks the really fluffy look of fungus). To treat it you will need professional advice.

476 Keeping the substrate clean helps to prevent infections

In addition to maintaining good water quality, it is also important to keep your substrate clean. There is some evidence that dirty gravels or those rich in decaying organic matter favor the outbreak of bacterial infections. This particularly applies to diseases such as mouth rot or mouth fungus, and the types of bacterial illness that can affect the mouths and barbels of catfish species.

477 Large external parasites are rarely seen in tropical fish

Many fishkeepers worry about introducing the large, external parasites, such as anchor worm *(Lernaea)*, fish lice *(Argulus)* and leeches, as these creatures look particularly alarming. However, they are really quite rare in tropical fish. If you do find your new stock is suffering from these, the adult parasites can be deliberately removed, and the free-swimming stages treated with prescription drugs or ultraviolet sterilization. Seek expert advice.

478 Internal parasites are difficult to treat

Fish can also suffer from internal parasites. There are few obvious symptoms, apart from lethargy and gradual emaciation. However, most of these infections are either caused by chronic stress, or are dependant on an intermediate invertebrate host, and cannot spread through aquariums. Thus they are rare in normal tanks. Treatment is difficult; if you have ruled out all other causes, simply give the fish the best care you can.

479 Dropsy is not a disease but a symptom

Dropsy is a swelling of the abdomen caused by excess fluid. Often, the pressure on the skin will cause the eyes to bulge, and the scales to stick out, giving the so-called pinecone effect. In fact, this is not a disease in itself, but a symptom of severe illness or organ failure. Although not usually infectious, isolate affected fish and seek expert help for treatment.

Tip 479 *Pineconelike scales are a symptom of dropsy.*

Tip 480 *A bala shark* (Balantiocheilus melanopterus) *with a cloudy eye.*

Eyes can tell you a lot about the health of your fish

Bulging eyes are one of the effects of dropsy. But eyes can tell you other things as well. A single bulging eye can be the result of violence from a tankmate, or infection, perhaps from poor conditions. A single blind eye is probably a simple deformity. Cloudy eyes indicate that the fish is unhappy with its water quality, while sunken eyes show that the fish is seriously ill.

481

Growths and tumors do not always need treatment

Like humans, fish can develop growths and tumors as a result of age, genetic or environmental factors. A veterinarian may be able to remove external lumps on a large fish, but there is a chance they could simply grow back. If you are certain that what your fish has is a benign growth and not a disease, there is no need to interfere unless the tumor prevents feeding or swimming.

Tip 481 *Tumors can develop for a variety of reasons.*

A general guide to diagnosing fish diseases

Sometimes, the symptoms of several diseases can resemble each other. As a (very general) guide to help you diagnose, bacterial diseases develop the fastest, in severe cases and from first signs to death is around 24 hours. Next are fungal infections, which almost always show "fluffy" blobs. Parasitic illnesses are generally slower, developing and spreading between individuals over a period of days.

Tip 483 *One red phantom tetra* (Megalamphodus sweglesi) *in the school is deformed.*

Do not breed from fish with congenital defects

Pet fish are sometimes a little inbred and can suffer from deformities. Missing gill covers, fins or eyes, kinked spines and irregular markings are common congenital defects. If a fish has been born with one or more of these, it is nothing to worry about, provided that it is not suffering. Simply ensure that you do not breed from these individuals.

Better Healthcare

484 Do not use medications as a preventative measure

Prevention is obviously better than cure, but do not use medications purely as a preventative measure without a very good reason. Many medications are slightly poisonous to fish, so never expose a healthy specimen needlessly to these chemicals. In addition, every time the disease-causing microorganisms come into contact with treatments, it gives them more opportunity to evolve resistance to them.

Tip 485 *A rapid response treatment kit should include: dechlorinator to treat tap water for emergency water changes, an antiparasite treatment, plus broad-spectrum antibacterial and fungus medications.*

Measuring cups and pipettes are included with treatments.

485 Keep an emergency medical kit handy for prompt action

When you see the first signs of an infection, you may need to start treatment right away. Depending on the location and opening hours of your nearest aquatic store, it could be a good idea to keep common medications in stock. Ensure you have a basic antiparasite, and an antibacterial and an antifungal treatment readily available, as well as plenty of dechlorinator for emergency water changes.

486 Some fish are sensitive to some medications

Certain species will not tolerate some medications. This is mainly a problem when using treatments against external parasites on some types of tetras, catfish and loaches, or cartilaginous fish such as stingrays. You may have to reduce the dosage. Simply observe the fish carefully when medicating, or use an alternative treatment. Seek expert advice from a professional or a book on fish diseases.

487 Keep a note of the capacity of your aquarium

You need to know the capacity of your aquarium when using medications. The volume is often clearly labeled on the tank, so make a note of it at the the time of purchase. Tip 1 shows how to calculate your tank's volume from its dimension. Gravel and ornaments will displace water; reduce the volume by approximately 10 percent or more, depending on your setup, to allow for this.

488 Additional aeration is beneficial during treatment

Some medications can reduce oxygen levels in the water, while others can irritate the fish's gills. Some do both. For these reasons it is important to ensure adequate aeration for the duration of a treatment. This is not usually a problem, but you may want to add an extra powerhead or air pump. Anything that creates surface movement or releases bubbles will do the trick. (See also Tip 97.)

Tip 488 *An airstone helps increase aeration.*

489 Always follow the directions when using medications

When adding medication for an illness, it may be necessary to treat the aquarium more than once to eradicate the disease. It is important to abide by the manufacturer's instructions, which usually involve a water change followed by re-dosing. Do not mix medications, except where this is specifically recommended. If you are treating more than one illness, treat the most serious one first.

490 The effect of carbon on tank medications

If your filter contains carbon, this will remove most medications from the water. Therefore, while using a treatment in the tank, it is important to take out the carbon part of your filter and run your equipment without it. Afterward, if you replace it with some new carbon, it will help to eliminate all traces of the medication from your setup.

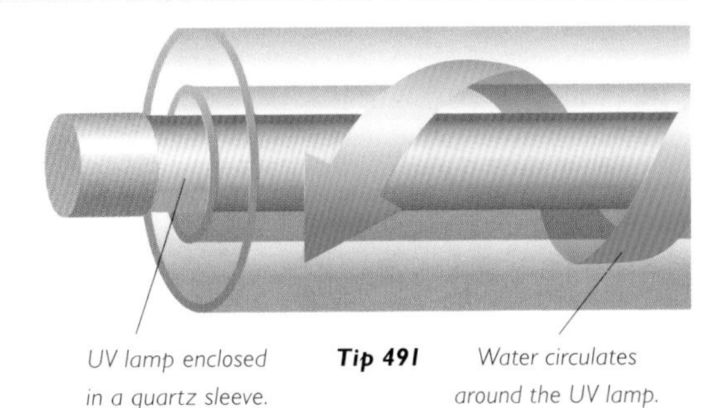

UV lamp enclosed in a quartz sleeve. **Tip 491** *Water circulates around the UV lamp.*

491 Try a UV sterilizer to eradicate parasites

If you need to treat a parasitic infection but your fish or invertebrates will not tolerate the medication, you can use an ultraviolet sterilizer. More commonly used in marine setups, an ultraviolet lamp is positioned outside the aquarium and the water is sterilized as it is pumped through it (see above). Parasites with a free-swimming stage, such as ich, can be eradicated this way. Ask your aquatic store for details.

Preparing medications

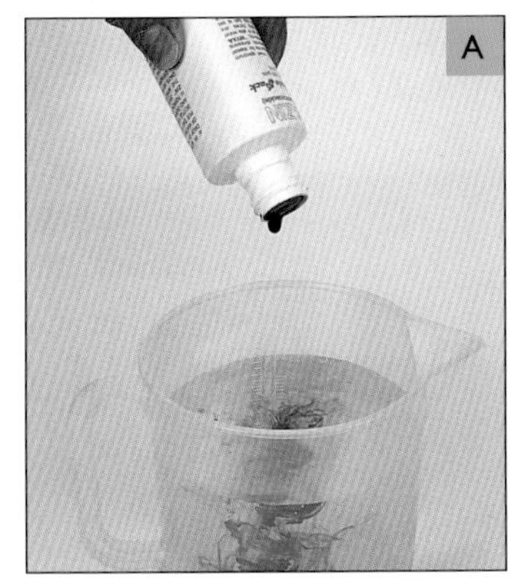

A *Calculate the correct amount of medication for your tank and add the solution to a plastic container of aquarium water.*

B *Thoroughly mix the medication into the water. By diluting it, you avoid producing localized spots of dangerously high concentrations.*

C *Gently introduce the diluted medication into the aquarium. Remember to keep any utensils solely for aquarium use.*

Better Healthcare

492 Can I use medications in an invertebrate tank?

Invertebrates and medications often do not mix. If you keep crabs, shrimps or ornamental snails and need to use a treatment, check the label first to see if it is safe for invertebrates. If not, move these animals to your quarantine aquarium. Inverts do not usually harbor fish diseases, so they can be safely returned to the main tank after the treatment has finished.

Tip 492 *Freshwater mussels may not tolerate tank medications.*

493 Can tank medications affect filter bacteria?

Some medications can damage your filter bacteria. Examples are copper, malachite green, methylene blue and some antibiotics. Often the effect is mild, but if treating for a bacterial disease keep an eye on ammonia and nitrite levels. If you do experience problems, feed lightly, consider using ammonia-neutralizers and be prepared to do some water changes (not forgetting to re-dose if necessary).

494 Can I catch any diseases from my fish?

There are very few diseases that fish can pass on to humans (most occur from eating fish rather than keeping them!). However, on rare occasions, people have been known to catch bacterial infections and possibly parasites from aquarium fish. Do not immerse open wounds in tank water (try wearing waterproof gloves), and do not siphon water from the aquarium with your mouth.

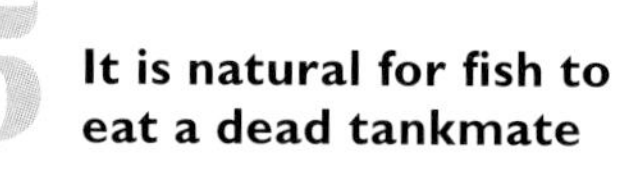

495 It is natural for fish to eat a dead tankmate

A common mistake is to see a fish chewing on the body of one of its companions, and automatically assume it must be responsible for the death. A dead fish is an easy meal and will be enjoyed by most species—foul play may not necessarily be involved. Check to see if the fish could have died from other causes before assuming aquatic homicide.

496 Remove dead fish from the tank as soon as possible

If you do lose a fish, remove the body as soon as you spot it. If it rots, it will place an extra burden on the filtration system and could raise ammonia levels. As importantly, if it remains in the tank the other fish will eat it, and some diseases can be passed on in this way, so it should be prevented as much as possible.

Tip 496 *Dead fish left in the aquarium become a health hazard.*

Tip 497 Angelfish (Pterophyllum scalare) may eat smaller tankmates.

497 Accounting for mysterious fish disappearances

Not all deaths are due to natural causes; although not common, some fish do view their tankmates as dinner rather than companions. This is worth considering if you lose several small fish in succession, but you cannot find any trace of the bodies (most true predators swallow whole), and the remaining individuals show no signs of ill-health. Check that all your species are compatible.

498 Try to establish the cause of fish deaths

Once a fish has been dead for more than 30 minutes, parasites will begin to leave the body, and the tissues begin to decompose. In addition, tankmates may nibble it, resulting in further damage. For these reasons it can be difficult to assess the cause of death accurately. Be sure to check that the remaining living fish are well and that all the tank conditions are in order.

499 Fish in the wild are also subject to disease

Remember that fish suffer from diseases in the wild as well. For example, in temperate regions, wild fish are known to experience outbreaks of ich, lice or bacterial infections on a regular basis. It is normal to encounter the occasional problem, but if you look after your fish as well as possible, the chances are that most of them will have a long and happy life.

500 Quarantine aquatic plants as well as fish

It is possible, although unlikely, for aquatic plants to transmit disease if they have come from a tank (or pond) containing fish. If you are concerned, the best option is to place new plants in your quarantine tank for a week or so. Quicker but less effective methods include dipping them in a broad-spectrum disease treatment, or simply giving them a gentle rinse under the tap.

Tip 500 Treat aquatic plants before adding them to the tank.

Credits

Unless otherwise stated, photographs have been taken by Geoff Rogers © Interpet Publishing.

The publishers would like to thank the following photographers for providing images, credited here by page number and position: (B) Bottom, (T) Top, (C) Center, (BL) Bottom left, etc.

Aqua Press (M-P & C Piednoir): 19, 57(T), 75(C), 76(R), 78, 81, 83, 84(T), 88(B), 107(CL), 110(BR), 123(B)

Interpet Publishing Ltd: 21(BR)

Jan-Eric Larsson-Rubenowitz: 31(CR), 47(B)

Photomax (Max Gibbs): 43(B), 103(T), 109

Reef One Ltd: 7(BL)

William A. Tomey: 12(TR)

Tropica: 34(T, Ole Pedersen), 39(BR)

Computer graphics by Phil Holmes and Stuart Watkinson © Interpet Publishing.

Acknowledgements

The publishers would like to thank the following for their help: Arcadia, Croydon, Surrey; Birds Fishing Tackle, Gt. Blakenham, Suffolk; Kesgrave Tropicals, Ipswich, Suffolk; Seapet Centre, Martlesham Heath, Suffolk; SLADAS, Leigh-on-Sea, Essex; Swallow Aquatics, Colchester, Essex, with special thanks to Lewis; Tropica Aquarium Plants A/S, Hjortshøj, Denmark.

Publisher's note

The information and recommendations in this book are given without guarantee on the part of the authors and publishers, who disclaim any liability with the use of this material.